8

Band 22 · August 1986

isotopenpraxis

AKADEMIE-VERLAG BERLIN

ISSN 0021-1915 Isotopenpraxis, Berlin 22 (1986) 8, 257–296

An unsere Autoren

Die Zeitschrift veröffentlicht Beiträge zu folgenden Themen: Kontrolle und Analyse radioaktiver und stabiler Isotope, Anwendung radioaktiver und stabiler Isotope in allen Gebieten der Forschung und Technik, der Industrie, Landwirtschaft und Medizin, Herstellung radioaktiver und stabiler Isotope sowie von markierten Verbindungen, kernelektronische Geräte für die Isotopenanwendung (Dickenmesser, Dichtemesser u. ä.) Detektoren für isotopentechnische Arbeiten und den Strahlenschutz in Isotopenlaboratorien, Meßtechnik und Strahlenschutz, Dosimetrie in Isotopenlaboratorien, kerntechnische Geräte für die Isotopenanwendung (Beton, Bestrahlungsanlagen u. ä.), Strahlenchemie, Radiochemie, ökonomische und perspektivische Fragen, Fragen der Weiterbildung.

1. Aufnahme von Manuskripten

Zur Veröffentlichung angenommen werden nur druckreife, noch nicht anderweitig publizierte Beiträge in deutscher, englischer oder russischer Sprache zum Themenbereich der Zeitschrift und zwar

- Originalarbeiten
- Kurze Mitteilungen
- Aus der Praxis
- Zusammenfassende Berichte
- Tagungsberichte
- Informationen.

Sind der Beitrag oder Teile daraus bereits veröffentlicht oder ist eine anderweitige Veröffentlichung vorgesehen, so muß dies der Redaktion mit genauer Angabe der Zweitveröffentlichung mitgeteilt werden. Mit der Annahme des Manuskriptes behält sich der Verlag das Recht der Vervielfältigung, Verbreitung und Übersetzung innerhalb der gesetzlichen Schutzfrist vor. Auszugsweiser Nachdruck oder photomechanische u. a. Wiedergabe sind an eine schriftliche Genehmigung des Verlages gebunden. Manuskripte, Zeichnungen und Photos von veröffentlichten Arbeiten werden nur auf den besonderen und bis zum Zeitpunkt der Veröffentlichung mitgeteilten Wunsch des Autors zurückgesandt.

Bei der unverlangten Einreichung von Manuskripten wird empfohlen, eine Befürwortung eines der Herausgeber oder Redaktionsausschußmitglieder beizufügen.

Manuskripte sind einzureichen bei der Redaktion, 7010 Leipzig, PSF 370.

2. Form der Manuskripte

Die Autoren werden gebeten:

2 Manuskripte, eines davon in Original-Maschinenschrift, einseitig und mit 2zeiligem Abstand einzureichen; gut leserliche Photokopien sind zulässig; auf dem 1. Blatt (unter Weglassung der Bezeichnung)

den Titel der Arbeit
Vorname-Initialen und Namen sämtlicher Autoren
gegebenenfalls die Institution, in der sie tätig sind
eine kurze Zusammenfassung der Arbeit von höchstens 15 Zeilen
Keywords in alphabetischer Reihenfolge und
als Fußnote die Anschrift der Institution anzugeben;

wenn möglich, Zusammenfassung, Titel und Autoren *englisch* (bei deutschem oder russischen Manuskript) bzw. *deutsch* (bei englischem Manuskript) 2fach beizufügen;

Literaturhinweise (fortlaufend numeriert), Fußnoten, Tabellen, Abbildungen (in 2—3facher Veröffentlichungsgröße), Abbildungsunterschriften (getrennt von den Abbildungen und Zeichenerklärungen, Quellenangaben, falls erforderlich, und gegebenenfalls mit einem Vermerk für besondere Wünsche bezüglich der Abbildungsgröße), private und Dienstanschrift der Verfasser mit akademischen Titeln und ausgeschriebenen Vornamen auf besonderen Blättern ans Ende des Manuskriptes zu stellen;

Wünsche für den Kolumnentitel (Seitenüberschrift) bekanntzugeben;
Absätze durch Einzüge von mindestens 3 Anschlägen kenntlich zu machen;
die Maßeinheiten des SI-Systems („Systeme International") zu verwenden;
Korrekturen im Manuskript unmittelbar an der zu korrigierenden Stelle anzubringen;

Formeln deutlich lesbar und möglichst groß einzutragen;
Farbabbildungen und nachträgliche Änderungen oder Ergänzungen in Abbildungen zu vermeiden;

hervorzuhebende Textstellen zu unterstreichen, und zwar halbfetten Satz mit einer geraden schwarzen Linie, kursiven Satz (Schrägschrift) mit einer blauen Wellenlinie (alle Eigennamen). Sperrungen gestrichelt;
besondere Buchstaben durch farbige Unterstreichung zu kennzeichnen, und zwar Formelzeichen in

Griechisch-Kursiv (α, β, ...) rot,
Kursiv (*A*, *B*, ...) blau
halbfetter Kursiv (**A**, **B**, ...) 2 Linien blau (Vektoren),
halbfetter Grotesk (**A**, **B**, ...) 2 Linien violett (Tensoren);

bei Literaturangaben auf Vollständigkeit zu achten. In nachstehender Reihenfolge müssen angegeben sein:

a) bei *Zeitschriften*: Vorname-Initialen und Namen sämtlicher Autoren, Titel der Zeitschrift (nach „Periodica Chimica" oder möglichst wenig gekürzt), Bandnummer (unterstrichen), Jahreszahl (in Klammern), Heftnummer (nur bei Zeitschriften, die jedes Heft mit S. 1 beginnen) sowie die Seitenzahl, auf der die zitierte Arbeit beginnt.

b) bei *Büchern*: Initialen der Vornamen und Name des Autors (Herausgeber werden durch den Zusatz „ed." gekennzeichnet), voll ausgeschriebener Titel des Buches in der Originalsprache, evtl. Bandnummer, voll ausgeschriebener Name des Verlages, Nummer der Auflage (außer der ersten), Erscheinungsort(e), Erscheinungsjahr (hier nicht in Klammern setzen!) und evtl. Kapitelnummer, Seitenzahl u. ä.

c) bei *Sammelbänden, Tagungsberichten*: Es soll zusätzlich der Titel des zitierten Beitrags angegeben werden. Die Seitenzahl muß unbedingt angegeben werden. Aus dem Zitat muß hervorgehen, ob es sich um einen Tagungssammelband oder um einen noch nicht veröffentlichten Vortrag handelt. Zwischen Tagungsort und Erscheinungsjahr des entsprechenden Tagungsberichts ist zu unterscheiden.

3. Korrekturen

Der Autor erhält zwei Umbruchsabzüge, von denen einer möglichst schnell korrigiert an die Redaktion zurückgereicht werden soll. Auf Wunsch des Autors kann die Korrektur auch von der Redaktion gelesen werden.

Durch nachträgliche Manuskriptänderungen verursachte Korrekturkosten gehen, sofern sie 10% der Satzkosten übersteigen, zu Lasten des Autors.

Der Autor erhält 50 Sonderdrucke kostenlos. Weitere Sonderdrucke können bis zur Rückgabe der Korrekturabzüge auf Kosten des Autors bei der Redaktion bestellt werden.

Bezugsmöglichkeiten

Bestellungen sind zu richten

- in der **DDR** an den Postzeitungsvertrieb unter Angabe der Kundennummer des Bestellenden oder an den AKADEMIE-VERLAG BERLIN, DDR-1086 Berlin, Leipziger Straße 3—4, PF-Nr. 1233;
- im **sozialistischen Ausland** an eine Buchhandlung für fremdsprachige Literatur oder an den zuständigen Postzeitungsvertrieb;
- in der **BRD und Berlin (West)** an eine Buchhandlung oder an die Auslieferungsstelle KUNST UND WISSEN, Erich Bieber OHG, Wilhelmstraße 4—6, D-7000 Stuttgart 1;
- in den **übrigen westeuropäischen Ländern** an eine Buchhandlung oder an die Auslieferungsstelle KUNST UND WISSEN, Erich Bieber GmbH, General Wille-Str. 4, CH-8002 Zürich;
- im **übrigen Ausland** an den Internationalen Buch- und Zeitschriftenhandel; den Buchexport Volkseigener Außenhandelsbetrieb der Deutschen Demokratischen Republik, DDR-7010 Leipzig, Postfach 160; oder an den AKADEMIE-VERLAG BERLIN, DDR-1086 Berlin, Leipziger Straße 3—4.

Zeitschrift „Isotopenpraxis"

Herausgeber: Zentralinstitut für Isotopen- und Strahlenforschung der Akademie der Wissenschaften der DDR.
Verlag: Akademie-Verlag Berlin, DDR-1086 Berlin, Leipziger Straße 3—4, PF-Nr. 1233; Fernruf: 223621 oder 2236229; Telex-Nr.: 114420;
Bank: Staatsbank der DDR, Berlin, Kto.-Nr.: 6836-26-20712.
Chefredakteur: Prof. Dr. H. Hübner, Leipzig.
Anschrift der Redaktion der Zeitschrift Isotopenpraxis: Prof. Dr. H. Hübner, DDR-7010 Leipzig, Karl-Marx-Platz, Hauptpostamt, PSF 370; Fernruf: 281018.
Veröffentlicht unter der Lizenznummer 1527 des Presseamtes beim Vorsitzenden des Ministerrates der Deutschen Demokratischen Republik.
Gesamtherstellung: VEB Druckerei „Thomas Müntzer", DDR-5820 Bad Langensalza.
Erscheinungsweise: Die Zeitschrift „Isotopenpraxis" erscheint monatlich. Die 12 Hefte einer Jahrganges bilden einen Band. Bezugspreis je Band 240,— DM zuzüglich Versandspesen; Preis je Heft 20,— DM. Der gültige Jahresbezugspreis für die DDR ist der Postzeitungsliste zu entnehmen.
Bestellnummer dieses Heftes: 1079/22/8.

AN (EDV) 12831
00900

Isotopenpraxis 22 (1986) 8, S. 257 bis 262

Ludwig Boltzmann — Pioneer of Atomistics and Evolution

W. Stiller

(Academy of Sciences of the GDR, Central Institute of Isotope and Radiation Research, Leipzig)[1])

At first a short introduction to Ludwig Boltzmann's life (1844—1906) and work is given. Some theoretical results of his work (H-theorem, classical Boltzmann statistics, Boltzmann's kinetic equation) are treated in some detail. His experimental work is briefly discussed. In addition Boltzmann's philosophical work is characterized. Finally the influence of Boltzmann's ideas on our time is investigated.

Zunächst wird eine kurze Einführung in Leben und Werk Ludwig Boltzmanns (1844—1906) gegeben. Einige theoretische Arbeiten Boltzmanns (H-Theorem, klassische Boltzmann-Statistik, Boltzmannsche Stoßgleichung) werden näher behandelt. Seine experimentellen Arbeiten werden kurz diskutiert. Des weiteren wird Boltzmanns philosophisches Werk charakterisiert. Abschließend wird der Einfluß von Boltzmanns Ideen auf unsere Zeit untersucht.

Keywords

atoms; gases; history; isotopes; statistical mechanics; thermodynamics

"I believe in molecules, and I can simply say: and yet they move."
(*Boltzmann* alluding to *Galileo Galilei* in 1897)

In our time when the fundamental notions of atomism are already introduced at school it is nearly unimaginable how distinguished scientists had to fight even about 1900 in order to make their ideas of the discontinuous structure of matter (atomistics) generally respected. One of the greatest them was *Ludwig Boltzmann*. The eightieth anniversary of his tragic death will be an opportunity to remember him.

Ludwig Boltzmann

Stations of his life and work

Ludwig Eduard Boltzmann was born on the 20th of February, 1844 at Vienna. His father served in the Imperial Internal Revenue Service. The family went to Wels, later to Linz, where *Ludwig* was very successful already at school. At 15 he lost his father but *Boltzmann*'s mother self-sacrificingly supported both his professional and his artistic education (studies in physics and mathematics at Vienna, Heidelberg and Berlin; piano lessons with *Anton Bruckner*). After taking his doctor's degree in 1866 *Boltzmann* worked as a lecturer with *Josef Stefan* at the University of Vienna and already in 1869 he became a professor of mathematical physics at Graz. That was precisely the period in which (in 1872) he derived his kinetic equation (today denoted as *Boltzmann* equation) describing the relaxation of the velocity distribution function of a dilute monatomic gas component to the Maxwellian distribution. From 1873 to 1876 *Boltzmann* taught mathematics at the University of Vienna again. In 1876, married to a young student of mathematics he returned to Graz as a professor of experimental physics (to succeed *Toepler* who went to Dresden). In this position *Boltzmann* carried out various experimental work, too (e.g. measurements of the dielectric constant in gases). Here at Graz he spent fourteen happy years: as a scientist, he realized his epoch-making research in statistical mechanics; as a family man, he was the father of two sons and three daughters. Step by step he became an excellent scientist well-known in Europe and America. In his town he was a Dean and temporarily Rector of Graz University; his country honoured him by conferring on him titles, such as "Regierungsrat", "Mitglied der Kaiserlichen Akademie" and "Wirklicher Hofrat". At that time his circle of students extended considerably; thus, *Arrhenius* and *Nernst* moved

[1]) Address: DDR-7050 Leipzig, Permoserstraße 15

to Graz to be taught theoretical physics by *Boltzmann*. In 1888 he was offered a chair in Berlin in order to take the vacant place of the late *Kirchhoff*. After longer negotiations [1] he finally refused the tempting Prussian offer, but with his later regret to have withdrawn from a valuable scientific sphere. Instead of *Boltzmann Planck* became *Kirchhoff*'s successor. In 1890, however, *Boltzmann* accepted an offer from Munich University as a "Professor of Theoretical Physics" — one of the first chairs in the field in Europe. In doing so he hoped to satisfy best his main interests in physics. *Pringsheim*, the mathematician, *Lommel*, the physicist, *Seeliger*, the astronomer, *Baeyer*, the chemist, and *Linde*, the engineer for cooling devices were his friends at the regular table of the "Hofbräuhaus". In 1893, his teacher and friend *Josef Stefan* died and *Boltzmann* returned to Vienna University as a Professor of Theoretical Physics. In 1895 here *Ernst Mach* — a person of great consequence — became a professor of philosophy. *Mach* was a prominent opponent of atomistics. Therefore *Boltzmann* was glad to be offered a chair of theoretical physics in Leipzig in 1900 due to an arrangement by the famous physical chemist *Wilhelm Ostwald*. Although making friends with him personally, *Boltzmann* was repelled by *Ostwald*'s "energetics" and anti-atomistics and did not feel at home spiritually in Leipzig. In 1902 he returned to Vienna again as his own successor on the vacant chair of theoretical physics. In addition, from 1903 he also gave lectures on the "method and general theory of science". Amongst other things in these lectures he taught Darwin's thoughts on evolution and extended them considerably to all living matter.

Overwork and internal consumption by continual violent discussions with his opponents (*Helm*, *Mach*, *Ostwald*) as well as deteriorated eyesight and his asthma made him depressive again and again. On the 5th of September 1906 he committed suicide at Duino near Triest.

Boltzmann's scientific work is comprehensive; a collection can be found e.g. with Broda [2]. It comprises the following books which all appeared in the publishing house of *J. A. Barth* at Leipzig:

"Vorlesungen über die Principe der Mechanik" (in 2 vols., 1897 and 1904), "Vorlesungen über Maxwells Theorie der Elektricität und des Lichts" (2 vols., 1891 and 1893, newly edited and introduced by *W. Kaiser*, Akad. Druck- und Verlagsanstalt Graz 1982), "Vorlesungen über Gastheorie" (2 vols., 1896 and 1898). In the same publishing house in 1905 *Boltzmann*'s "Populäre Schriften" appeared comprising 23 longer papers. A new insignificantly shortened edition is due to *Broda* [3] and was published by *Vieweg*, Braunschweig in 1979. It is the one we want to refer to in what follows. In 1907 *Boltzmann*'s fundamental contribution "Kinetische Theorie der Materie" (coauthor, *J. Nabl*) was published in the "Encyclopädie der mathematischen Wissenschaften" (vol. V, part 8) by the publishing house of Teubner in Leipzig. Finally in 1909 *Boltzmann*'s disciple and later successor at Vienna University *Friedrich Hasenöhrl* [4] edited his master's 139 collected short original papers entitled "Wissenschaftliche Abhandlungen von Ludwig Boltzmann" (3 vols.).

What can we learn from this brief description of Boltzmanns' life and work in the present time?

First, *Boltzmann* was above all (but by no means exclusively) a theoretical physicist with universal abilities and interests.

Second, in spite of his passionate statements in favour of atomistics *Boltzmann* was a representative of the epoch of the classical physics of the 19th century.

Third, *Boltzmann* was a philosopher (evolutionist), too.

Fourth, notwithstanding his stays in Munich (4 years) and Leipzig (2 years) *Boltzmann* was above all an Austrian physicist.

Finally it is noteworthy that the tragedy of his relatively early death is still increased by the fact that *Einstein* (1905) and *v. Smoluchowski* (1906) theoretically and *Perrin* (1908) experimentally proved the unambiguously discrete structure of matter (existence of atoms and molecules) by studying the Brownian motion of particles.

In what follows these simple statements will be considered in more detail in order to come to a just assessment of *Boltzmann*'s personality, work and lasting importance for science.

Boltzmann as a (theoretical) physicist

Boltzmann's work in theoretical physics was extraordinarily many-sided as can be seen already from the enumeration of his books and articles. This statement can be qualified in that he was mainly interested in a further development of theory, especially in a molecular-kinetic foundation of thermodynamics. Stimulated by the mechanical theory of heat set up by *Maxwell* and *Clausius* he put himself the question how a dilute gas can reach the stationary Maxwellian equilibrium state. The equation describing the relaxation of an arbitrary non-equilibrium velocity distribution function $f(\boldsymbol{r}, \boldsymbol{v}, t)$ depending on the particle position $\boldsymbol{r}$ and time t to its equilibrium form $f^{(\mathrm{eq})}(\boldsymbol{v})$ is nowadays called *Boltzmann*('s) (kinetic) equation. In modern notation it can be formulated as follows (comp. [5] with [6]):

$$\frac{\partial f}{\partial t} = -\boldsymbol{v} \cdot \frac{\partial f}{\partial \boldsymbol{r}} - \boldsymbol{a} \cdot \frac{\partial f}{\partial \boldsymbol{v}} + \int \mathrm{d}\boldsymbol{v}_1 \times \int \mathrm{d}\Omega \, g\sigma(g, \Theta) \, (f' f_1' - f f_1) \tag{1}$$

(for details of the derivation see, e.g., [7]). The quantity $f(\boldsymbol{r}, \boldsymbol{v}, t) \, \mathrm{d}\boldsymbol{r} \, \mathrm{d}\boldsymbol{v}$ gives the number of monatomic molecules in the position-velocity element $\mathrm{d}\boldsymbol{r} \, \mathrm{d}\boldsymbol{v}$ of the so-called μ-space forming a 6-dimensional space. In this manner *Boltzmann* intended to balance all the possible physical effects (e.g. collisions, external fields) which could be have an influence on the non-equilibrium distribution function f of the velocities $\boldsymbol{v}$ of the molecules at the position $\boldsymbol{r}$ at time t. The quantity $S(f) = -\boldsymbol{v} \cdot \frac{\partial f}{\partial \boldsymbol{r}} - \boldsymbol{a} \cdot \frac{\partial f}{\partial \boldsymbol{v}}$ describes the change of f in time as a result of motions of independent particles (molecules, atoms) in μ-space. $S(f)$ is named "streaming term". The term $\int \mathrm{d}\boldsymbol{v}_1 \int \mathrm{d}\Omega \ldots$ is called collision integral. It describes the change of f as a result of binary collisions. *Boltzmann*'s famous Stoßzahlansatz (collision number hypothesis) $f(\boldsymbol{v}) \, f(\boldsymbol{v}_1)$ gives the probability (density) for a collisional event of two particles with velocities $\boldsymbol{v}$ and $\boldsymbol{v}_1$, respectively, leading to particles with new velocities $\neq \boldsymbol{v}$ after the collision. In other words, such (direct) collisions deplete the reservoir of particles with velocities $\boldsymbol{v}$. Therefore the term $f f_1$ in the collision integral is balanced with a negative sign. Conversely, however, collisions of particles which before the collision had velocities $\neq \boldsymbol{v}$ can lead to an enrichment of $\boldsymbol{v}$-particles after a binary collision ("restituting collisions"). This effect characterized by the primed quantities in eq. (1) gives rise to a positive sign. The collision integral extended to all the particle velocities $\boldsymbol{v}_1$ and the solid angle Ω describes the physics of (elastic) binary collisions comprised in the important quantities g (= magnitude $|\boldsymbol{v} - \boldsymbol{v}_1|$ of the relative velocity) and $\sigma(g, \Theta)$ (differential cross section, depending on g and the scattering angle Θ). Mathematically speaking *Boltzmann*'s kinetic equation is a non-linear integro-differential equation for f the solution of which is difficult to find in general. In what most remarkable way does *Boltzmann* solve the problem in question? Given at time $t = 0$ an arbitrary initial distribution function $f(\boldsymbol{r}, \boldsymbol{v}, 0)$, then *Boltzmann* defines a function H (later denoted as *Boltzmann*'s H-function):

$$H(t) = \int \mathrm{d}\boldsymbol{r} \int \mathrm{d}\boldsymbol{v}\, f \ln f \tag{2}$$

For this function *Boltzmann* can prove the following property: For any given initial distributions $H(t)$ will decrease monotonically in time until an equilibrium state is reached which can be characterized by the functional equation:

$$f f_1 = f' f_1' \tag{3}$$

Eq. (3) gives solutions in the general form:

$$f(\boldsymbol{r}, \boldsymbol{v}, t) = A \exp\left(-\frac{\beta m}{2} (\boldsymbol{v} - \boldsymbol{u})^2\right) \tag{4}$$

Here the quantities A, β, and $\boldsymbol{u}$ are functions depending in general on $\boldsymbol{r}$ and t. For many but by no means all the external potentials $U(\boldsymbol{r})$[2]) it can be inferred from eqs. (1) and (4) the following solution:

$$f = A \exp\left(-\beta\left[\frac{m}{2}\boldsymbol{v}^2 + U(\boldsymbol{r})\right]\right) \tag{5}$$

Neglecting external forces $\big(U(\boldsymbol{r}) = 0\big)$ and considering the conditions of conservation of the number of particles and total energy of the gas the Maxwellian distribution function of the velocities $\boldsymbol{v}$ of a perfect gas is obtained:

$$f(\boldsymbol{v}) = n\left(\frac{m}{2\pi kT}\right)^{3/2} \exp\left(-\frac{m\boldsymbol{v}^2}{2kT}\right) \tag{6}$$

n number density, T absolute temperature, m mass of an individual particle of the gas.

Boltzmann's H-function including its tendency to decrease monotonically to an equilibrium state for arbitrary initial distributions (content of *Boltzmann*'s H-theorem) is identical with a microscopic (statistical) foundation of the second law of thermodynamics: If we construct a function $S(t)$ proportional to $(-H)$ then according to the H-theorem S is a monotonically increasing function during the relaxation process up to the point where its maximum is attained. In other words, the function $S(t)$ fulfils the property of the entropy of the gas.

Serious critism was raised against the H-theorem already in *Boltzmann*'s days. One point is *Loschmidt*'s Umkehreinwand (1876). The argument refers to the following contradiction. The equations of classical mechanics describing the (classical) motions of molecules are reversible in time as can be simply seen by a change of t into $(-t)$ in *Newton*'s equations of motion without varying them. In contrast, the H-theorem describes an irreversible behaviour of a gas system as a whole. *Boltzmann* himself was the first who could weaken the contradiction in the sense that the Stoßzahlansatz only holds for an average number of collisions but for an overwhelming majority of diluted gaseous systems such a statistical approach (neglecting fluctuations) is reasonable. It is worth mentioning that for theorists it has been an attractive business up to our days to derive *Boltzmann*-like kinetic equations and to look for the point at which irreversibility has been "smuggled in" somehow. Another interesting theoretical approach reconciling reversible mechanics with irreversible thermodynamics is made by *Prigogine* and coworkers, who introduce irreversible elements already into mechanics (see e.g. *Prigogine*'s formulation of a microscopic entropy operator [8]).

A second objection against *Boltzmann*'s H-theorem was *Zermelo*'s Wiederkehreinwand (1896) based on *Poincaré*'s theorem: any mechanical system moving in finite regions of phase space will practically return to its origin after a certain time — a fact which cannot be reconciled with irreversible behaviour. Recently *Rice*, *Freed* and *Light* [9] pointed out that *Zermelo*'s argument is unjustified because *Boltzmann*'s H-theorem is in connection with the distribution function whereas *Poincaré*'s theorem refers to the trajectories immediately.

Boltzmann did not only describe the approach to equilibrium, but he also revealed the atomic (statistical) structure of entropy — perhaps his greatest theoretical success. Due to *Planck* [10] this can be reflected in the following way. On the one hand, it is known from thermodynamics that entropy (as introduced by *Clausius*) is an extensive quantity, i.e. the total entropy of two gaseous subsystems 1 and 2 (independent of each other, with individual entropies S_1 and S_2) is

$$S = S_1 + S_2 \tag{7}$$

On the other hand, the total probability W of finding the two independent subsystems in states with the probabilities W_1 and W_2, respectively, is

$$W = W_1 \cdot W_2 \tag{8}$$

Summing up the statements (7) and (8) the result is

$$S = k \ln W \qquad (k \text{ constant}) \tag{9}$$

This is the famous *Boltzmann* relation (*Einstein* called it *Boltzmann*'s principle), which by the way, decorates *Boltzmann*'s gravestone at the Wiener Zentralfriedhof. The constant k in eq. (9) like that in (6) was never numerically calculated by *Boltzmann* himself; this was done by *Planck* [10]. Nevertheless today k is named *Boltzmann*'s constant with $k = R/N_L$ (R universal gas constant, N_L *Loschmidt*'s number[3])). The central quantity W is named thermodynamic probability. *Boltzmann* worked out the prescription how to combine it with atomistics: W is the number of microstates which can form a macrostate described e.g. by such phenomenological quantities like temperature or pressure of the gas. In order to assess the number of thinkable microstates making up a macrostate *Boltzmann* used the mathematical branch of combinatorial analysis. He assigned every molecule (or atom) a number and counted the possibilities of distributing N given molecules over Z cells of phase space. The cells are assumed to be big enough for being occupied by the microscopic particles without mutual hindrance. In the framework of *Boltzmann*'s statistics no more assumptions on the properties of the particles are necessary than their denumerability and distinguishability. Obviously *Boltzmann* did not want

[2]) *Boltzmann* even studied such exceptions in detail [6].

[3]) *Josef Loschmidt* (1821–1895) — an older Vienna colleague and close friend of *Boltzmann* — tried to estimate the size of a molecule.

to anticipate at this point the future development of atomic physics because he felt that soon details on the atomic structure would be brought to light by using spectral analysis. The decisive fact for *Boltzmann* was only the assertion that matter consists of discrete particles.

It is interesting to note that reputable physicists did not believe in atoms and molecules at that time, whereas most of the chemists used the concept of the atom as a useful working hypothesis. The latter was certainly caused by the successes of *Mendeleev*'s periodic table and of organic chemistry with its ideas of molecular structures. To *Boltzmann* atoms were real in so much as they seemed him to be relatively stable particles which were in rapid agitation. *Boltzmann*, however, did not believe in their indivisibility as required by the Greek atomists *Leucippus* and *Democritus*.

Besides his H-theorem and statistical atomistics *Boltzmann* succeeded in elaborating numerous further physical concepts and laws, e.g., the derivation of *Stefan*'s radiation law, the equipartition theorem, the collision theory of elastic cylinders, the theory of *van der Waal*'s equation, the theory of the *Hall* effect. It should be mentioned also his calculations of thermoelectricity and his fundamental papers on diffusion, internal friction and heat conductivity.

During the period when *Boltzmann* was a professor of experimental physics at Graz he achieved important results on this topic, too. According to *Schaefer* [11], above all *Boltzmann*'s investigations (together with *Toepler*) on the motion of air in sounding pipes are worth mentioning. Later he measured the dielectric constants of solids (e.g. hard rubber, sulphur, paraffin, colophony). With the same measurements in gases he could corroborate *Maxwell*'s relation of the electromagnetic theory of light. *Boltzmann* informed *Helmholtz* in Berlin in a remarkable letter about his successful finding [11]. Finally should be mentioned "*Boltzmann*'s Doppelspiegel" (double-mirror) — an extension of *Fresnel*'s mirror to the invisible region of the electromagnetic spectrum.

Boltzmann as an evolutionist

The physicist *Boltzmann* gained national and international acknowledgement already during his lifetime. His philosophical activity as an evolutionist, however, is by far less known. *Broda* emphatically referred to this in his writings (see e.g. [2, 12]). *Boltzmann*'s engagement in philosophy can primarily be connected with his universal understanding of science as such. Further aspects, however, can and must be added. Thus obviously a considerable part of his philosophical views resulted from the physical world of statistical mechanics itself which in describing mechanical evolution deals with 10^{23} Newtonian equations of motion. But *Boltzmann* felt that for a formulation of thermodynamic evolution the dynamics of the 10^{23} particles need not be known even if this were possible. Furthermore, *Boltzmann* was strongly influenced by *Charles Darwin*'s theory of evolution. He even denoted the 19th century as the century of *Darwin* [13]. Finally from 1903 *Boltzmann* engaged in philosophy by giving lectures on the "method and general theory of science" at Vienna University as *Mach*'s successor.

Boltzmann's opinion on evolution in science can be best rendered by some personal remarks. From *Darwin*'s thoughts on evolution he concluded for the process of photosynthesis in plants [14]: "The general struggle of the organisms is . . . not a struggle for the elements — the elements of all organisms are abundant in air, water and soil — nor for energy, which in the form of heat, unfortunately unchangeable, is contained in all substances, but a struggle for the entropy, which becomes available in the transition of the energy from the hot Sun to the cold Earth. To exploit this transition as far as possible the plants spread out the immeasurable area of their leaves and force solar energy in a way as yet unexplained, before it sinks down to the temperature level of the surface of the Earth, to carry out chemical syntheses of which one has no inkling as yet in our laboratories. The products of this chemical kitchen are the object of the struggles of the animals."

On the evolution of the human brain *Boltzmann* said [15]: "The brain is considered by us as the instrument, the organ for the production of world pictures, which because of their great utility for the maintenance of the species according to *Darwin*'s theory developed to particular perfection in man, as in the giraffe the neck, and in the stork the beak developed to unusual length."

What about *Boltzmann*'s views on the theory of cognition ? In his opinion our notions are instruments consciously or unconsciously produced by man in order to express experience in a concentrated way. The notions can fail if the range of experience is too widely exceeded. On the laws of thinking *Boltzmann* said [16]: "I am convinced that the laws of thinking have been formed by a better and better fitting of combinations of ideas within us (which we design for the objects) to the combinations of objects. All the combination rules which led to contradictions with experience were rejected whereas the true ones have decidedly been conserved. The latter have consistently been transmitted from generation to generation. In this way we have come to believe in such combinations as if they were inborn customs of thinking."

Influence of Boltzmann's ideas on our time

It is not so easy to evaluate *Boltzmann*'s universal influence on our time. Nevertheless the author of this short memorial paper is of opinion that in characterizing *Boltzmann*'s importance 80 years after his death the following aspects should not be forgotten:

At his time *Boltzmann* was an outstanding scientist and lecturer. As regards the latter enthusiastic reports of his students (e.g. *Lise Meitner* or *Paul Ehrenfest*) spoke about "kristallklare" lectures he had given in overcrowded lecture-rooms.

Owing to his lectures on electrodynamics *Boltzmann* (like *Helmholtz* and *Stefan*) can be credited with helping to rapidly propagate *Maxwell*'s electromagnetic theory over the European continent; doubtless such a development here led to a fast progress in physics and electrical engineering including the production of modern devices of communication.

With respect to the mechanical theory of heat *Boltzmann* ingeniously developed the ideas of *Krönig*, *Clausius* and *Maxwell*.

Boltzmann's statistics is to a certain extent the close of classical physics. Simultaneously, as a result of its far-reaching atomistic concept *Boltzmann*'s statistics provides a fruitful soil for quantum physics (e.g. *Planck*'s radiation law, *Bose-Einstein-* and *Fermi-Dirac* statistics, *Bohr*'s and

Sommerfeld's atomic theory and quantum mechanics). For this reason *Alfred Kastler* denoted *Boltzmann* as the "father" and *Planck* as the "mother" of quantum theory [17]. By introducing the calculus of probability into theoretical physics *Boltzmann* succeeded in a microscopic (statistical) description of macroscopic quantities like entropy which had originally been formulated phenomenologically. Parallel to this *Boltzmann* contributed to the second law of thermodynamics with a deep analysis. In this way he became a pioneer of important developments in solid-state physics (*Nernst*'s third heat law; *Einstein*'s explanation of the drop of specific heats of solids with decreasing temperatures).

Influences of *Boltzmann*'s ideas extending into our time can be found in information theory and also in transport theory in which *Boltzmann*'s kinetic equation is of fundamental importance [18, 19]. Thus every year e.g. in the fields of gas dynamics [19, 20], plasma physics of (low pressure) discharges [21, 22], transport theory of neutrons [23], solid state physics [24], and chemical reaction kinetics [25, 26] there appear hundreds of papers starting from *Boltzmann*-like equations such as the *Waldmann-Snider*-equation, *Wang-Chang-Uhlenbeck* equation, *Vlasov*-equation, *Lenard-Balescu*-equation, *Enskog*-equation, *Fokker-Planck*-equation, and so on. The quantities of change under consideration have changed, of course: While *Boltzmann* primarily investigated velocity distributions functions of simple gases, today translational energy distributions and transport coefficients also of complex polyatomic molecules [27], electrons [28—30] and ions [31—33], radicals [33] and hot atoms [34] are of interest. In regard to charged particles their kinetics in time-dependent external electric and magnetic fields is under consideration [35]. Certainly the methods of solving kinetic equations have also been developed further. We would like to mention here at least *Carleman*'s solution of the initial problem, the methods of the linearized *Boltzmann* equation, the *Chapman-Enskog* and the *Grad* expansion [19], the solution theory of *Bobylev*, *Krook* and *Wu* [36]. In overcoming practical difficulties of solving the *Boltzmann* equation — especially, if several concurring kinds of collisions are under consideration — often Monte-Carlo methods are of importance [37]. By the way, it is also of interest that the *Boltzmann* equation can be generalized to higher velocities by using the theory of relativity as could be shown by *Lichnerowicz* and *Marrot* [38]. A summary in the field of the *Boltzmann* equation and related problems was given in comprehensive papers of an international symposium on "100 Years Boltzmann Equation" held in Vienna in 1972 [39]. Finally mention should be made of the efforts to generalize *Boltzmann*'s H-theorem ($-\mathrm{d}H/\mathrm{d}t \geqq 0$) to a so-called super-$H$-theorem with the property $(-1)^n \times \mathrm{d}^nH/\mathrm{d}t^n \geqq 0$ ($n = 1, 2, \ldots$), in order to single out the H-function as the correct non-equilibrium entropy from a large class of functions which would also given the correct value of the equilibrium entropy [40].

Together with *Willard Gibbs* (1839—1903) of Yale College, New Haven, whose ensemble statistics he highly esteemed, *Ludwig Boltzmann* also has to be considered as the founder of statistical mechanics. Considering *Boltzmann*'s work as a whole *Broda* came to the conclusion that besides *Isaac Newton* and *James Clerk Maxwell Ludwig Boltzmann* was the most important theoretical physicist of the classical epoch.

As a fervent defender of atomistics *Boltzmann* faced the opposition of distinguished intellectual adversaries, as, for example, all his life the positivist *Ernst Mach*, the "Energetiker" *Wilhelm Ostwald* (who could only be convinced of atomistics in 1908), the French physicist *Pierre Duhem* as well as *Max Planck* in his younger years.

As a physicist working mainly theoretically *Boltzmann* strove for a close relationship to philosophy and recommended such a procedure also to his colleagues. As an evolutionist *Boltzmann* propagated *Darwin*'s theory of evolution and applied the idea of evolution — in the widest interpretation of the notion "mechanics" — to all living matter up to human thinking. Thus he can also be considered as one of the founders of (theoretical) biophysics.

Finally the readers of "Isotopenpraxis" should pay attention to another important "distant effect' of *Boltzmann*'s work which is felt even today. That is, applying his famous relation $S = k \ln W$ with its statistics for W to different isotopic species of the same chemical element (which can be done in a convenient manner via the partition function formalism) we arrive at the statistical theory of isotope systems. A description of physical and chemical isotope effects is thus the fruit of the isotope statistics as can be learnt from numerous contemporary papers in the field in this journal. ... "Nothing is more practical than a good theory" was a popular (but also obligatory) saying of *Boltzmann* addressed to all his students and friends . . .

References

[1] *H. J. Treder*, Große Physiker und ihre Probleme, Berlin: Akademie-Verlag 1982, p. 105

[2] *E. Broda*, Ludwig Boltzmann, Mensch — Physiker — Philosoph, Vienna: Franz Deuticke 1955

[3] *E. Broda* (ed.), Ludwig Boltzmann, Populäre Schriften, Braunschweig-Wiesbaden: Vieweg 1979

[4] *F. Hasenöhrl* (ed.), Wissenschaftliche Abhandlungen von Ludwig Boltzmann, Leipzig: J. A. Barth 1909

[5] L. Boltzmann, Wiener Berichte 66 (1872) 275 or [4], vol. 1, p. 316

[6] *G. E. Uhlenbeck*, *G. W. Ford*, Lectures in Statistical Mechanics, Providence/Rhode Island: American Mathematical Society Publ. 1963

[7] *P. E. Zweifel*, The Boltzmann Equation and Its Properties. In: *C. Cercignani* (ed.), Kinetic Theories and the Boltzmann Equation, Berlin — Heidelberg — New York — Tokyo: Springer-Verlag 1984 (Lecture Notes in Mathematics; 1048)

[8] *I. Prigogine*, From Being to Becoming. Time and Complexity in Physical Sciences, München, Zürich: R. Piper 1979

[9] *S. Rice*, *K. F. Freed*, *J. C. Light* (eds.), Statistical Mechanics: New Concepts, New Problems, New Applications, Chicago 1972

[10] *M. Planck*, Wissenschaftliche Selbstbiographie, 5th ed., Leipzig: J. A. Barth 1970

[11] *C. Schaefer*, Naturwiss. 33 (1946) 33

[12] *E. Broda*, Ethics and Evolution in Boltzmann's and Einstein's Thought. In: *M. Kageyama* et al. (ed.), Science and Scientists, Dordrecht — Boston: Reidel 1981

[13] see [3] pp. 26—46

[14] *E. Broda*, Comp. Biochem. Physiol. 67B (1980) 373

[15] see [3] pp. 94—119

[16] see [3], pp. 170—190

[17] *E. Broda*, The Intellectual Quadrangle: Mach-Boltzmann-Planck-Einstein, colloquium CERN 81-10, Geneva 1981

[18] *C. Cercignani*, Theory and Application of the Boltzmann Equation, Edinburgh: Scottish Academic Press, and New York: Elsevier 1975

[19] *J. H. Ferziger*, *H. G. Kaper*, Mathematical Theory of Transport Processes in Gases, Amsterdam — London; North-Holland 1972

[20] *B. Shizgal*, *J. M. Fitzpatrick*, J. chem. Phys. 63 (1975) 131, 138

[21] *A. R. Hochstim* (ed.), Kinetic Processes in Gases and Plasmas, New York: Academic 1969

[22] *L. S. Polak*, Teoretičeskaja i prikladnaja plazmochimija, Moskva: Nauka 1975

[23] *I. Kuššer*, Acta Physica Austriaca, Suppl. 10 (1973) 491

[24] *R. Kubo*, Acta Physica Austriaca, Suppl. 10 (1973) 301

[25] *J. Ross, P. Mazur*, J. chem. Phys. **35** (1961) 19
I. Prigogine, E. Xhrouet, Physica **15** (1949) 913
R. Der, R. Haberlandt, A. Merkel, H. J. Czerwon, Chem. Physics **53** (1980) 427
[26] *J. Troe*, Ber. Bunsenges. phys. Chemie **72** (1968) 908
R. Haberlandt, Nova Acta Leopoldina NF **49**, Nr. 233 (1979) 27
[27] *L. Waldmann*, Acta Physica Austriaca, Suppl. **10** (1973) 223
[28] *J. Wilhelm, R. Winkler*, J. de physique C7 (1979) 251
[29] *M. Capitelli, C. Gorse, J. Wilhelm, R. Winkler*, Ann. Physik (Leipzig) 7. Folge, Bd. **41** (1984) 119, Chem. Phys. **79** (1983) 1;
J. Wilhelm, R. Winkler, Contrib. Plasmaphysics **20** (1980) 225
[30] *R. Schmidt, W. Stiller*, Contrib. Plasmaphysics **23** (1983) 41, Chem. Phys. Lett. **118** (1985) 564
[31] *G. L. Braglia*, Contrib. Plasmaphysics **20** (1980) 147
[32] *L. Viehland, S. L. Lin, E. A. Mason*, Chem. Phys. **54** (1981) 341;
S. L. Lin, R. E. Robson, E. A. Mason, J. chem. Phys. **71** (1979) 3483
[33] *W. Stiller, R. Schmidt*, Z. physik. Chemie (Leipzig) **261** (1980) 271;
W. Stiller, doctoral thesis B, Akademie der Wissenschaften der DDR, Berlin 1978
[34] *B. Shizgal* J. chem. Phys. **74** (1981) 1401; **72** (1980) 3156;
B. Shizgal, J. M. Fitzpatrick, J. chem. Phys. **72** (1980) 3143
[35] *W. Stiller, G. Vojta*, Phys. Lett. **27A** (1968) 478, Physica **48** (1969) 216; Z. Naturforsch. **24a** (1969) 545, 555
[36] *M. R. Hoare*, Adv. chem. Physics **56** (1984) 1–140
[37] *K. Koura*, J. chem. Phys. **82** (1985) 2566
[38] *S. R. de Groot*, Acta Physica Austriaca, Suppl. **10** (1973) 529
[39] *E. G. D. Cohen, W. Thirring* (eds.), The Boltzmann Equation. Theory and Applications, Wien – New York: Springer-Verlag 1973 (Proceed. Int. Symp. "100 Years Boltzmann Equation", 4th–8th Sept. 1972, Vienna); refs. [23], [24], [27], [38] are also included in [39].
[40] *M. H. Ernst*, J. statistical Physics **34** (1984) 1001; For mathematical refinements of the notion of entropy (see e.g.): *A. Uhlmann*, Rep. Math. Phys. **1** (1970) 147; for *H*-theorems of the *Fokker-Planck*-equation: *B. Crell, T. de Paly, A. Uhlmann*, Wiss. Z. d. Karl-Marx-Univ. (Leipzig) **27** (1978) 229

Isotopenpraxis **22** (1986) 8, S. 262 bis 269

Originalarbeiten

Eignung verschiedener ^{15}N-Verbindungen als Tracer für Körpermarkierung und für Untersuchungen des Protein- und Aminosäuremetabolismus

K. Krawielitzki, R. Schadereit, U. Herrmann

(Forschungszentrum für Tierproduktion Dummerstorf-Rostock der Akademie der Landwirtschaftswissenschaften der DDR, Bereich Tierernährung „Oskar Kellner") [1]

In einem Versuch mit wachsenden Ratten werden die Resorption, die Retention und die intermediäre Verteilung des ^{15}N nach zehntägiger Verabfolgung von vier verschiedenen ^{15}N-Tracern: DL-(^{15}N)-Methionin, L-(^{15}N)-Lysin, (^{15}N)-Glycin und ($^{15}NH_4)_2SO_4$ untersucht und die Übertragungsprozesse der markierten Aminogruppe (Transaminierung, Desaminierung und Aminierung sowie andere chemische Umwandlungen zwischen den Aminosäuren) bestimmt. Die Ergebnisse werden diskutiert und daraus Schlußfolgerungen für die Eignung der Substanzen zur Markierung von Organismen bzw. als Tracer für die Bestimmung von Parametern des Protein- und Aminosäurenstoffwechsels abgeleitet.

In an experiment with growing rats the resorption, the incorporation and the intermediate fate of ^{15}N were estimated using four different ^{15}N-tracers: DL-(^{15}N)-methionine, L-(^{15}N)-lysine, (^{15}N)-glycine and ($^{15}N_4)_2SO_4$. Transformation of the labelled amino group by transamination, by desamination and amination as well as by other chemical processes of amino acids will be checked. The results are discussed and conclusions are drawn for the use of ^{15}N-substances as a labelling substance of organisms respectively as a tracer for estimating parameters of the whole body protein and amino acid metabolism.

Keywords

amino acids; ammonium sulfates; glycine; lysine; methionine; nitrogen 15; rats; tracer techniques

1. Einleitung

Die Anwendung markierter Aminosäuren und anderer Tracerverbindungen zur Untersuchung des intermediären Protein- und Aminosäurenmetabolismus stellt seit vielen Jahren sowohl in der Humanmedizin als auch in der Tierernährung eine gängige Methode dar. In Abhängigkeit von der Zielstellung der Versuche, vom biologischen Objekt und von der Art des eingesetzten Tracers wurden unterschiedliche Versuchs- und Auswertungsmethoden entwickelt, die nicht immer identische Werte ergeben [1], aber doch meist zu brauchbaren und vergleichbaren Aussagen und Ergebnissen führen [1, 2].

In der Praxis werden vielfach ^{14}C- oder ^{3}H-markierte Aminosäuren als Tracer verwendet, die durch einmalige Injektion oder durch Dauerinfusion verabfolgt werden [3, 4, 5; Literaturzusammenfassung 6].

Neben radioaktiv markierten Verbindungen werden in zunehmendem Maße auch stabilisotop markierte Verbindungen als Tracer für Gesamtkörperstoffwechseluntersuchungen eingesetzt (vorwiegend ^{15}N-markierte Substanzen). Sie weisen gegenüber radioaktiven Tracern eine Reihe von Vorteilen auf [7]:

1. Wegen des Fehlens von radioaktiver Strahlung eignen sich ^{15}N-Tracer auch zur Untersuchung des menschlichen Metabolismus.
2. Es erfolgt keine Autoradiolyse der markierten Verbindungen.

[1]) Anschrift: DDR-2500 Rostock, Justus-von-Liebig-Weg, PF 27-01/02

3. Der Protein- und Aminosäurenstoffwechsel wird durch die markierte Aminogruppe besser repräsentiert als durch die Kohlenstoffkette, da ^{15}N nicht in andere Körperbausteine, wie Kohlenhydrate und Lipide, eingebaut werden kann.

Demgegenüber steht jedoch der Nachteil einer wesentlich geringeren Nachweisempfindlichkeit bei stabilen Isotopen und somit die Notwendigkeit, ^{15}N-Verbindungen in größeren Mengen zu applizieren. Bei emissionsspektrometrischer Messung der ^{15}N-Häufigkeit ist eine maximale Isotopenverdünnung von 1:3000 noch erfaßbar, bei hochauflösenden Massenspektrometern ist diese noch um eine bis zwei Zehnerpotenzen darunter meßbar [8], bei Radionukliden dagegen beträgt sie $1:10^6$ bis $1:10^8$. Die Nachweisempfindlichkeit für ^{15}N ist jedoch für die meisten Fragestellungen ausreichend und daher ist die ^{15}N-Tracertechnik gegenwärtig vor allem im humanmedizinischen Bereich international die gebräuchlichste Methode [1]. Die ^{15}N-Tracertechnik wurde z. B. bei *Waterlow* u. a. [2] sowie bei *Hartig* u. a. [9] ausführlich beschrieben.

Nach *Faust* u. a. [1] wird von den ^{15}N-Tracerverbindungen das [^{15}N]-Glycin mit nahezu 50% aller Fälle weitaus am häufigsten angewendet. Danach folgen Harnstoff, mit ca. 25% Anteil, markierte Proteine, [^{15}N]-Lysin und [^{15}N]-Leucin. Alle anderen Tracerverbindungen werden nur vereinzelt angewendet.

Von den Tracertechniken rangiert die Pulsmarkierung an erster Stelle, gefolgt von der Infusionsmethode [1]. Für bestimmte Fragestellungen, z. B. Studium der endogenen Sekretion, der Proteinverdaulichkeit, wird auch eine Vormarkierung des Organismus über eine längere Markierungsperiode vorgenommen. Für diese Zwecke wurden biosynthetische Proteine von Hefe [10, 11, 12], Eiern [12, 13], Milch und Getreide [10, 15] sowie ^{15}N-markierte Ammoniumsalze [16, 17, 18, 19, 20] eingesetzt, bei Wiederkäuern auch ^{15}N-Harnstoff [21, 22] bzw. ^{15}N-Ammoniumcarbonat [23, 24, 25].

Die Eignung einzeln eingesetzter AS und speziell des (^{15}N)-Glycins als Tracer für die Untersuchung des Proteinstoffwechsels wurde wiederholt in Zweifel gezogen [z. B. 2, 26, 27, 28, 29].

Bei der Markierung des Organismus mit ^{15}N-Ammoniumsalzen wurde festgestellt, daß überwiegend die nichtessentiellen Aminosäuren markiert wurden und daher die gemessenen Stoffwechselparameter im wesentlichen den Umsatz dieser Aminosäuren widerspiegeln [30].

In den vorliegenden Versuchen sollte die Resorption, die Retention und die Verteilung des ^{15}N-Isotops nach oraler Applikation unterschiedlicher Tracerverbindungen im intermediären Stoffwechsel untersucht und der Markierungsgrad der einzelnen Aminosäuren des Körperproteins geprüft werden, um daraus Schlußfolgerungen für die Eignung der Substanz zur Markierung des Organismus bzw. als Tracersubstanz für den Protein- und Aminosäurenstoffwechsel abzuleiten.

2. Material und Methode

Zwölf männliche Wistar-Ratten (Anfangs-Lebendmasse 63 g) wurden in vier Gruppen eingeteilt und erhielten zehn Tage lang eine Futtermischung aus N-freiem Teilfutter, Mineral- und Wirkstoffmischungen sowie folgenden Proteinträgern:

Gruppe 1: Kasein + DL(^{15}N)-Methionin (mit 95% Atom-% ^{15}N-Exzeß)
Gruppe 2: Weizengluten + L(^{15}N)-Lysin · 2 HCl (mit 95 Atom-% ^{15}N-Exzeß)
Gruppe 3: Volleipulver + (^{15}N)-Glycin (mit 95 Atom-% ^{15}N-Exzeß)
Gruppe 4: Volleipulver + $(^{15}NH_4)_2SO_4$ (mit 54,1 Atom-% ^{15}N-Exzeß)

Die Zusammensetzung der vier Rationen ist in Tab. 1 dargestellt. Die Futtereinnahme betrug einheitlich 8,70 g Originalsubstanz (= 7,84 g Trockensubstanz). Die Futterrationen waren isoenergetisch. Die Aminosäuren Methionin und Lysin wurden auf einen Anteil von 4,0 bzw. 5,0% vom Rohproteingehalt der jeweiligen Ration ergänzt. Als Proteinträger in den Gruppen 3 und 4 wurde hier Vollei gewählt, da es einen hohen Anteil an essentiellen Aminosäuren enthält und damit eine gute Verwertung der nichtessentiellen Supplemente (^{15}N)-Glycin bzw. $(^{15}NH_4)_2SO_4$ zu erwarten ist. Die markierten Substanzen wurden den Rationen täglich zugewogen.

Nach zehntägiger Fütterung mit dem markierten Futter wurden die Tiere getötet, der Gastrointestinaltrakt entfernt und in den homogenisierten Resttierkörpern der N-Gehalt bestimmt.

Weiterhin wurden nach Hydrolyse der Tierkörperhomogenate die Gehalte und die ^{15}N-Konzentrationen der einzelnen Aminosäuren gemessen. Die N-Bestimmung erfolgte nach dem *Kjeldahl*-Prinzip, die quantitative Aminosäurenanalyse und die präparative Trennung der Aminosäuren wurden säulenchromatographisch mit dem AS-Analysator LC 4010 durchgeführt und die ^{15}N-Bestimmung emissionsspektrometrisch mit dem Isonitromat 5201. Die Methodik der präparativen Aminosäurentrennung und der anschließenden ^{15}N-Analyse der AS-Fraktionen ist in einer früheren Publikation beschrieben worden [31].

3. Ergebnisse und Diskussion

3.1. *N- und ^{15}N-Einnahme*

Die N- und die ^{15}N-Einnahme sowie die Lebendmasseentwicklung der Versuchstiere während der 10tägigen Mar-

Tab. 1. Futterzusammensetzung der vier Rationen in g/d

	Gruppe 1 DL-(^{15}N)-Methionin (95 Atom-% ^{15}N)	Gruppe 2 L-(^{15}N)-Lysin (95 Atom-% ^{15}N)	Gruppe 3 (^{15}N)-Glycin (95 Atom-% ^{15}N)	Gruppe 4 $(^{15}NH_4)_2SO_4$ (54,1 Atom-% ^{15}N)
Kasein	1,30	–	–	–
Weizengluten	–	2,34	–	–
Volleipulver	–	–	2,36	2,36
N-freie Teilmischung	2,11	2,11	2,11	2,11
Maisstärke	5,26	4,17	4,15	4,16
^{15}N-Tracerverbindung	0,030	0,083	0,080	0,067
Originalsubstanz	8,70	8,70	8,70	8,70

Tab. 2. N-Einnahme, ^{15}N-Einnahme und Lebendmasseentwicklung der Versuchstiere

	Gruppe 1 DL-(^{15}N)-Methionin	Gruppe 2 L-(^{15}N)-Lysin	Gruppe 3 (^{15}N-Glycin)	Gruppe 4 ($^{15}NH_4)_2SO_4$
	$n = 3$	$n = 2$	$n = 3$	$n = 3$
N-Einnahme				
aus Protein [mg/d]	167	278	181	181
aus Tracer [mg/d]	3,0	10,9	15,77	15,0
^{15}*N-Einnahme* [mg/d]	2,85	5,20	15,00	8,12
Lebendmasse				
Versuchsanfang [g]	63,0 ± 2,5	63,0 ± 0,5	63,3 ± 2,3	63,8 ± 0,8
Versuchsende [g]	94,8 ± 7,8	93,8 ± 3,2	104,4 ± 5,3	103,2 ± 3,4
Lebendmassezunahme [g/d]	3,2 ± 0,65	3,1 ± 0,28	4,1 ± 0,61	3,9 ± 0,30

kierungsperiode sind in Tab. 2 zusammengestellt. Die mittlere Lebendmasse über die zehntägige Versuchszeit betrug für alle Tiere 81,4 ± 3,7 g, die mittlere Lebendmassezunahme 3,62 ± 0,63 g/d. Futterreste traten nur bei einem Tier der Gruppe 2 auf; da dieses Tier auch in Bezug auf Endmasse und Lebendmassezunahme deutlich abfiel, wurde es nicht in die Auswertung einbezogen. Die Bestimmung der ^{15}N-Gehalte in den einzelnen Fraktionen und in den Aminosäuren erfolgte jeweils nur von zwei Tieren jeder Gruppe.

Die Lebendmassezunahmen der Versuchstiere zeigten keine signifikanten Unterschiede. Insgesamt kann auf Grund der vollständigen Futteraufnahme und der kontinuierlichen und gleichmäßigen Lebendmassezunahme bei allen Versuchstieren auf eine normale Entwicklung der Ratten geschlossen werden.

3.2. *Resorption und Retention des ^{15}N-Tracers*

Aus der Literatur ist bekannt, daß bei oraler Verabfolgung synthetischer ^{15}N-markierter Aminosäuren oder ^{15}N-markierter Ammoniumsalze mit einer vollständigen Resorption des ^{15}N gerechnet werden kann [32–35].

Im vorliegenden Versuch wurde die fäkale ^{15}N-Ausscheidung nicht erfaßt. In früheren Versuchen bei Verfütterung von ^{15}N-markiertem Glycin an Ratten wurden maximal 4% [36] und an Schweinen 3,5 bis 8,6% des verabfolgten ^{15}N im Kot wiedergefunden. Analoge Ergebnisse wurden auch bei oraler Verabfolgung von (^{15}N)-Lysin an Schweine erzielt [37]. Da der bei diesen Versuchen im Kot ausgeschiedene ^{15}N vorwiegend in der TCE-fällbaren Fraktion vorliegt, kann angenommen werden, daß diese Reste keine unvollständig resorbierten ^{15}N-Tracer darstellen, sondern von endogenen Proteinen stammen.

Von dem resorbierten ^{15}N wurden unterschiedlich hohe Mengen ausgeschieden. Der Verbleib und die Verteilung des aufgenommenen ^{15}N sind in Tab. 3 dargestellt. Die N-Gehalte der Tierkörper waren für alle vier Gruppen annähernd gleich; der erreichbare ^{15}N-Überschuß ist aber für die einzelnen ^{15}N-Substanzen sehr unterschiedlich. Für Methionin und Lysin ist die maximale Supplementierung durch den Bedarf an diesen Aminosäuren begrenzt, wenn keine Imbalanz auftreten soll. Die tägliche Zufuhr an ^{15}N über diese Aminosäuren betrug nur 2,85 bzw. 5,2 mg und der ^{15}N-Überschuß im Tierkörper 0,44 Atom-% ^{15}N' für die Methionin- und 1,18 Atom-% ^{15}N' für die Lysingruppe.

Für (^{15}N)-Glycin und (^{15}N)-Ammoniumsulfat besteht diese Restriktion im Prinzip nicht. Für Gruppe 3 wurde eine Einnahme von 15 mg ^{15}N/Tag angestrebt und bei Gruppe 4 die N-äquivalente Menge ^{15}N-markiertes Ammoniumsulfat verabfolgt. Auf Grund des unterschiedlichen Markierungsgrades beider ^{15}N-Verbindungen besteht keine ^{15}N-Äquivalenz.

Der im Tierkörper eingebaute ^{15}N-Anteil differiert für die einzelnen Gruppen deutlich. Für (^{15}N)-Methionin (Gruppe 1) ist die Retention mit 34% der Einnahme am geringsten, für (^{15}N)-Lysin (Gruppe 2) mit 57% am höchsten. Beide Werte sind jedoch für essentielle Aminosäuren relativ niedrig. In früheren Versuchen erreichten wir bei lysindefiziter Fütterung der Schweine (^{15}N)-Lysin Verwertungskoeffizienten von 94% [37]. Der hohe oxydative Abbau des Methionins resultiert sicher aus seinen Funktionen im Stoffwechsel (Methylgruppendonator).

Eine relativ hohe ^{15}N-Einbaurate (ca. 50% der Einnahme) wurde auch bei Verabfolgung von ^{15}N-Glycin beobachtet (Gruppe 3).

Das steht in Übereinstimmung mit Versuchen verschiedener Autoren an Ratten und Schweinen, wobei in Ab-

Tab. 3. Übersicht über die ^{15}N-Einnahme und -Verteilung nach zehntägiger Markierungsperiode

	Gruppe 1 DL-(^{15}N)-Methionin		Gruppe 2 L-(^{15}N)-Lysin		Gruppe 3 (^{15}N)-Glycin		Gruppe 4 ($^{15}NH_4)_2SO_4$	
	Tier 2	Tier 3	Tier 4	Tier 5	Tier 7	Tier 9	Tier 10	Tier 11
Tierkörper								
N-Gehalt [mg]	2441	2725	2519	2513	2716	2482	2584	2740
^{15}N-Überschuß								
Atom-% ^{15}N'	0,42	0,46	1,13	1,22	2,87	3,06	0,95	1,23
mg ^{15}N'	9,03	10,35	28,5	30,7	72,9	77,9	28,9	33,7
^{15}*N-Einnahme*								
[mg ^{15}N']	28,5	28,5	52,0	52,0	150,0	150,0	81,2	81,2
^{15}N-Resorption [%]	100	100	100	100	100	100	100	100
davon:								
^{15}N-Retention (%)	31,7	36,3	54,8	59,0	48,6	52,0	35,6	41,5
^{15}N-Einbau in AS (%)	19,4	23,5	52,1	52,6	31,5	33,1	23,3	24,2
^{15}N-Einbau in sonst. Verbindungen (%)	12,3	12,8	2,7	6,4	17,1	18,9	12,3	17,2

hängigkeit vom Proteingehalt der Ration sogar Verwertungskoeffizienten von 65 bis 80% bzw. 75 bis 90% erreicht wurden [18, 36, 38].

Bei Gruppe 3 wurde nach 10tägiger Markierung (mit täglichen Gaben von 15 mg ^{15}N/Tier) ein Überschuß im Tierkörper von ca. 3,0 Atom-% bestimmt. Für ^{15}N-Ammoniumsulfat (Gruppe 4) betrug die ^{15}N-Einbaurate ca. 40% und der Markierungsgrad im Rattenkörper ca. 1,1 Atom-%. *Hernandez* u. a. [16] sowie *Simon* u. a. [17] erhielten in analogen Versuchen mit ausgewachsenen Ratten unter Erhaltungsbedingungen ebenfalls eine Ausnutzung von ca. 30% des ^{15}N aus Ammoniumsalzen. In weiteren Versuchen wurden $^{15}NH_4$-Citrat [17] sowie ein Gemisch von $^{15}NH_4Cl$ und $^{15}NH_4$-Acetat [19, 20] als Markierungssubstanz für Ratten bzw. Schweine verwendet, wobei in Abhängigkeit vom Proteinniveau und vom Rohfasergehalt ^{15}N-Einbauraten von 10% bis 47% erreicht wurden.

Aus allen diesen Resultaten lassen sich für die Resorption und die Retention von ^{15}N bei verschiedenen ^{15}N-Tracern folgende Schlußfolgerungen ziehen:

1. ^{15}N-Ammoniumsalze und ^{15}N-markierte Aminosäuren werden rasch und vollständig resorbiert.
2. Die ^{15}N-Markierung des Organismus ist mit ^{15}N-Ammoniumsalzen, mit ^{15}N-markierten Aminosäuren und auch mit ^{15}N-markierten Proteinen möglich. ^{15}N-Ammoniumsalze ergeben relativ geringe Einbauraten für ^{15}N, sind aber wegen ihres geringeren Preises gegenüber ^{15}N-markierten Aminosäuren ökonomisch günstiger.
3. Das Anion des als Tracer eingesetzten ^{15}N-Ammoniumsalzes scheint von untergeordneter Bedeutung zu sein, kann aber bei höheren Konzentrationen zu physiologischen Nebenwirkungen führen. Am wenigsten sind diese bei Salzen organischer Säuren, wie $^{15}NH_4$-Citrat, und bei $^{15}NH_4$-Acetat zu befürchten.

3.3. *Intermediäre Verteilung des inkorporierten ^{15}N*

Wie die Tab. 3 zeigt, wird nur ein Teil des aufgenommenen ^{15}N inkorporiert (vgl. auch [37]).

Von der ^{15}N-Einnahme waren nach zehntägiger Markierungsperiode letztlich bei (^{15}N)-Methionin 21,5%, bei (^{15}N)-Lysin 52,4%, bei (^{15}N)-Glycin 32,2% und bei $(^{15}NH_4)_2SO_4$ 23,8% in Aminosäuren eingebaut. Der ^{15}N-Einbau in Aminosäuren im Verhältnis zur ^{15}N-Retention lag bei Methionin (Gruppe 1), bei Glycin (Gruppe 3) und $(^{15}NH_4)_2SO_4$ (Gruppe 4) zwischen 62 und 65%. Nur bei (^{15}N)-Lysin (Gruppe 2) war der Anteil wesentlich höher (89 bzw. 95% des inkorporierten ^{15}N). Etwa ein Drittel des von (^{15}N)-Methionin, (^{15}N)-Glycin und $(^{15}NH_4)_2SO_4$ stammenden markierten Stickstoffs repräsentiert also andere Substanzen als Aminosäuren.

Bei Gabe von (^{15}N)-Ammoniumsalzen verläuft der ^{15}N-Einbau in die Aminosäuren über die Bildung von (^{15}N)-Ammoniak, das durch Aminierung an α-Ketoglutarsäure angelagert wird. Die Glutaminsäure bildet dann den Ausgangsstoff für die ^{15}N-Markierung weiterer Aminosäuren durch Transaminierung, das Glutamin für die Markierung von Purinen, Pyrimidinen, Aminozucker u. a. Stoffwechselprodukten über verschiedene Synthesewege [39, 40].

Im folgenden soll die Verteilung des markierten Stickstoffs auf die einzelnen Aminosäuren untersucht werden. In Tab. 4 sind die Aminosäure-Gehaltswerte nach der ^{15}N-Überschuß der einzelnen Aminosäuren im Tierkörper nach Verabreichung der vier verschiedenen ^{15}N-Tracer angegeben. Die Aminosäurezusammensetzung im Tierkörper ist unabhängig vom angebotenen Futter relativ konstant, wie die geringen Standardabweichungen ausweisen. Analoge Befunde für die Konstanz des Aminosäurespektrums, unabhängig von der Ration, liegen für Ratten auch von *Pellet* und *Kaba* [41] sowie von *Köhler* u. a. [42] vor.

In der ^{15}N-Anreicherung weisen die als Tracer verwendeten Aminosäuren im Tierkörper bei weitem die jeweils höchsten Werte auf, d. h. die verabfolgte ^{15}N-markierte Aminosäure wurde in großen Mengen unverändert inkorporiert. Daneben zeigen aber auch die anderen Aminosäuren in nahezu allen Fällen deutlich erhöhte ^{15}N-Konzentrationen. Der markierte Stickstoff kann auf drei verschiedene Arten von der Traceraminosäure auf andere

Tab. 4. Aminosäuregehalt und ^{15}N-Überschuß in Aminosäuren von Ganzkörperhomogenaten nach zehntägiger Markierungsperiode mit ^{15}N-Aminosäuren bzw. mit ^{15}N-Ammoniumsulfat

AS	AS-Gehalt	^{15}N-Überschuß (Atom-%)			
	g/16 g N	Gruppe 1 DL-(^{15}N)-Methionin	Gruppe 2 L-(^{15}N)-Lysin	Gruppe 3 (^{15}N)-Glycin	Gruppe 4 $(^{15}NH_4)_2SO_4$
Glu	13,13 ± 0,39	0,45	0,26	1,72	1,91
Ala	6,39 ± 0,35	0,32	0,23	1,45	1,28
Asp	8,68 ± 0,15	0,36	0,22	1,05	1,20
Arg	6,44 ± 0,19	0,25	0,14	0,85	1,12
Met	1,55 ± 0,06	*4,93*	0,37	0,96	1,11
Leu	7,63 ± 0,19	0,29	0,17	1,40	0,98
Val	4,73 ± 0,32	0,27	0,20	0,81	0,97
Ile	3,69 ± 0,11	0,25	0,19	0,81	0,95
Ser	4,65 ± 0,16	0,24	0,17	(7,57)	0,87
Pro	5,89 ± 0,25	0,12	0,09	0,57	0,77
Gly	8,11 ± 0,26	0,22	0,09	*9,59*	0,75
Cys	1,60 ± 0,08	0,39	0,09	(3,13)	0,42
His	2,62 ± 0,11	0,10	0,17	0,24	0,26
Phe	4,04 ± 0,11	0,15	0,11	0,23	0,34
Lys	6,83 ± 0,19	0,05	*11,61*	0,56	0,22
Thr	4,23 ± 0,15	0,03	0,04	0,15	0,24
Tyr	3,33 ± 0,12	nicht bestimmt		nicht bestimmt	

Σ AS 93,54 ± 2,84
davon: Σ ess. AS 46,69 ± 1,28, Σ nichtess. AS 46,85 ± 1,56

Aminosäuren übertragen werden:

1. durch Transaminierungsprozesse,
2. durch direkte Umwandlungsprozesse zwischen den Aminosäuren
3. durch oxydative Desaminierung und anschließende Reaminierung der gebildeten α-Ketosäuren.

Daneben kann auch eine mikrobiologische Synthese von Aminosäuren bzw. Proteinen aus ^{15}N-markierten Verbindungen im Darmtrakt erfolgen. Dieser Weg spielt jedoch nur bei Wiederkäuern eine Rolle [43], bei Nichtwiederkäuern ist er für den Eiweißstoffwechsel praktisch bedeutungslos, da die Aminosäuren nicht resorbiert werden [37].

Transaminierung. Der Hauptanteil der Übertragung der ^{15}N-Markierung auf andere Aminosäuren erfolgt durch Transaminierungsprozesse. Transaminierungen sind reversible Reaktionen mit geringer Energietönung, bei denen die Übertragung der NH_2-Gruppe von einer Aminosäure auf die andere ohne intermediäre Freisetzung von NH_3 erfolgt und bei welchen die meisten Aminosäuren ineinander überführt werden können. Sie wurden durch Aminotransferasen katalysiert, verlaufen streng stereospezifisch und stellen einen wichtigen Weg für die Anpassung des Aminosäurenstoffwechsels an die jeweiligen Erfordernisse dar. Die Gesamtmenge der Aminosäuren bleibt dabei unverändert; die Transaminierung bedeutet also keine „Neusynthese" von Aminosäuren. Hauptschaltstelle für die Transaminierung sind Glutaminsäure und Asparaginsäure, die neben den durch direkte Umwandlung gebildeten Aminosäuren in allen vier Gruppen den höchsten ^{15}N-Überschuß aufweisen. (Der relativ hohe ^{15}N-Überschuß im Methionin bei Gruppe 2 kann nicht erklärt werden!) Es folgen Alanin und Glycin. Den geringsten ^{15}N-Überschuß zeigen Threonin, Phenylalanin und Lysin. Dies sind die Aminosäuren, die aus sterischen Gründen kaum oder gar nicht an der Transaminierung teilnehmen. Die relative Markierungshöhe der jeweiligen Aminosäure ist damit ein Maß für den Grad der Beteiligung dieser Aminosäure an der Transaminierung. Der relativ hohe ^{15}N-Einbau charakterisiert die enorme Dynamik dieser Austauschprozesse für die (markierte) Aminogruppe. Aus den Werten in Tab. 4 kann auf eine mittlere Umsetzungsrate von 20 – 50% des Poolumsatzes geschlossen werden. Dieser rasche Austausch der Aminogruppe zwischen den Aminosäuren führt über eine längere Versuchszeit zu einem gewissen Ausgleich im Markierungsgrad der einzelnen Aminosäuren im AS-Pool.

Umwandlungsprozesse zwischen Aminosäuren. Direkte Umwandlungsprozesse spielen eine untergeordnete Rolle und haben nur in speziellen Fällen Bedeutung, da essentielle Aminosäuren nicht ineinander umgewandelt werden können. So kann Methionin bis zu einem gewissen Grade irreversibel zu Cystein [44] und Glycin in der Leber reversibel zu Serin [44] umgeformt werden. Bei der Umwandlung von Methionin zu Cystein wird die markierte Aminogruppe nicht mit übertragen.

Die Umwandlung von Serin in Glycin erfolgt mit Hilfe der C_1-Übertragung durch Tetrahydrofolsäure. Umgekehrt kann Serin durch C_1-Übertragung aufgebaut werden. In beiden Fällen bleibt die C-N-Bindung zur markierten Aminogruppe erhalten. Daraus resultiert der relativ hohe Markierungsgrad des Serins im Tierkörperprotein nach Verabfolgung von (^{15}N)-Glycin (Gruppe 3).

Aus Methionin wird bei der Transmethylierung durch Abgabe seiner labilen Methylgruppe das Homocystein, das mit dem Serin die SH-Gruppe gegen die OH-Gruppe austauschen kann. Dabei entstehen Cystein und Homoserin und als Zwischenprodukt die Additionsverbindung Cystathionin.

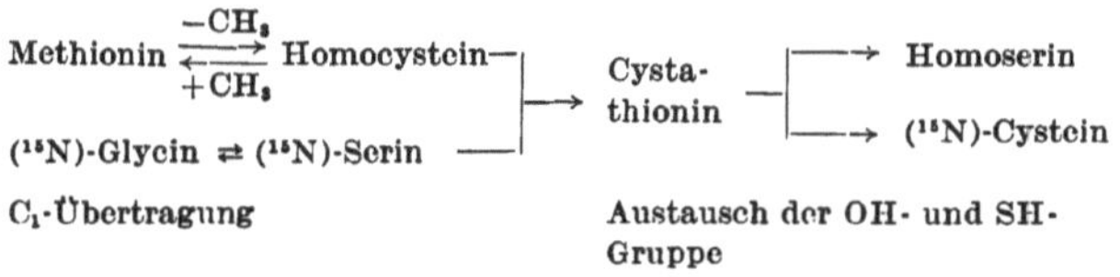

Auf diese Weise kann durch Aminosäuren-Umwandlung auch ^{15}N-markiertes Cystin aus (^{15}N)-Glycin über (^{15}N)-Serin gebildet werden. Die Intensität des Umwandlungsprozesses von Glycin und Serin und weiter zu Cystin wird durch die Höhe des Markierungsgrades dieser beiden Aminosäuren im Vergleich zu den übrigen Aminosäuren in Gruppe 3 deutlich. Der erreichte ^{15}N-Überschuß beträgt beim Serin 7,57 Atom-% (= 79% des ^{15}N-Überschusses von Glycin) und beim Cystin 3,13 Atom-% (= 33% des ^{15}N-Überschusses von Glycin). Für die übrigen Aminosäuren liegt der erreichte ^{15}N-Markierungsgrad dagegen zwischen 18% (Glutaminsäure) und 1,6% (Threonin) des Glycinwertes.

Zu analogen Ergebnissen kamen auch *Conway* u. a. [45]. Sie fanden nach (^{15}N)-Glycinverabfolgung an Patienten die in Tab. 5 dargestellten Ergebnisse. *Stein* [44] spricht sogar von einem konstanten ^{15}N-Verhältnis zwischen (^{15}N)-Glycin und (^{15}N)-Serin. Der hohe ^{15}N-Überschuß des Serins bei Verabfolgung von (^{15}N)-Glycin scheint damit hinreichend erklärt zu sein.

Tab. 5. ^{15}N-Anreicherung in verschiedenen Aminosäuren des Blutserums von Patienten nach Verabfolgung von (^{15}N)-Glycin bei drei unterschiedlichen Proteinniveaus *Conway* u. a. 1980]

Aminosäure	^{15}N-Anreicherung im Verhältnis zum Glycin
Glycin	100 %
Serin	56,8 ± 4,9 %
Glutaminsäure + Glutamin	14,6 ± 2,6 %
Alanin	8,6 ± 2,6 %
Isoleucin	7,0 ± 5,0 %
Leucin	6,5 ± 2,5 %
Valin	4,1 ± 1,3 %
Ornithin	6,5 ± 4,1 %
Prolin	3,9 ± 2,2 %

Auffallend in Tab. 4 ist weiterhin der relativ hohe Markierungsgrad von Arginin im Vergleich mit anderen Aminosäuren speziell bei Gruppe 3 und 4. Nach *Hofmann* [46] ist die α-Aminogruppe des Arginins nicht an der Transaminierung beteiligt. Es ist also zu vermuten, daß die ^{15}N-Markierung dieser Aminosäure aus der Arginylgruppe stammt und durch direkte Umsetzungsprozesse über den Ornithinzyklus in das Argininmolekül gelangt ist. Auch das Prolin kann auf diesem Wege ^{15}N einbauen. Gestützt wird diese Vermutung durch die Tatsache, daß bei Gabe von ^{15}N-Ammoniumsalzen die Markierung von Arginin wesentlich höher ist als bei Verabfolgung von ^{15}N-Traceraminosäuren. Die ^{15}N-Markierung des Arginins erfolgt also überwiegend oder gänzlich durch direkte Umwandlungsprozesse.

Desaminierung und Aminierung. Die dritte Möglichkeit der Übertragung von ^{15}N auf andere Aminosäuren stellt die oxydative Desaminierung und reduktive Reaminierung von Aminosäuren dar. Über diesen reversiblen Weg, der im Gegensatz zur Transaminierung über freies NH_3 führt und die Fixierung oder Bildung von freiem Ammoniak ermöglicht, ist auch eine Einführung von ^{15}N aus ^{15}N-

Ammoniumsalzen (Gruppe 4) in die Aminosäuren sowie die Umwandlung von D-Aminosäuren in L-Aminosäuren (Gruppe 1) möglich. Die Umwandlung von D-Methionin in L-Methionin erfolgt über oxydative Desaminierung mit D-2-Aminosäureoxydase und anschließender Aminierung des Ketomethionins zum L-Methionin [46]. Bei dieser Umlagerungsreaktion wird die markierte NH_2-Gruppe des Methionins gegen eine unmarkierte ausgewechselt und das freigesetzte markierte NH_3 kan auf dem Harnwege ausgeschieden oder über weitere Stoffwechselreaktionen in andere Aminosäuren bzw. N-Verbindungen eingebaut werden. Daraus resultiert auch der für eine essentielle Aminosäure geringe Verwertungskoeffizient beim DL(^{15}N)-Methionin. Außerdem spielt sicher die Funktion des Methionins als Methylgruppendonator im intermediären Stoffwechsel eine Rolle.

Die Reihenfolge im Markierungsgrad der einzelnen Aminosäuren nach ^{15}N-Ammoniumsalzgabe ist praktisch die gleiche wie bei Verabfolgung von ^{15}N-Traceraminosäuren nach Eliminierung der Traceraminosäure sowie der Aminosäuren, die durch direkte chemische Umwandlung aus der Traceraminosäure gebildet werden können. An der Spitze liegen Glutaminsäure, Asparaginsäure und Alanin, am Ende Threonin, Lysin, Histidin und Phenylalanin.

Ein direkter Vergleich der ^{15}N-Markierung einzelner Aminosäuren zwischen den vier Gruppen ist wegen der Unterschiede in der ^{15}N-Einnahme und in der ^{15}N-Verwertung in dieser Form nicht möglich. Die Unterschiede im Markierungsgrad der Glutaminsäure als der Schlüsselverbindung für die Transaminierung der Traceraminosäure sind aber beträchtlich.

$$\text{Gruppe 1 [(}^{15}\text{N)-Methionin]:} \quad \frac{\text{Atom-\% }^{15}\text{N (Glu)}}{\text{Atom-\% }^{15}\text{N (Met)}} = 0{,}09$$

$$\text{Gruppe 2 [(}^{15}\text{N)-Lysin]:} \quad \frac{\text{Atom-\% }^{15}\text{N (Glu)}}{\text{Atom-\% }^{15}\text{N (Lys)}} = 0{,}02$$

$$\text{Gruppe 3 [(}^{15}\text{N)-Glycin]:} \quad \frac{\text{Atom-\% }^{15}\text{N (Glu)}}{\text{Atom-\% }^{15}\text{N (Gly)}} = 0{,}18$$

Demnach wird die ^{15}N-markierte Aminogruppe des Glycins sehr intensiv, die des Lysins dagegen kaum in die Transaminierungsprozesse einbezogen. Methionin nimmt eine Mittelstellung ein.

Eine bessere Vergleichsmöglichkeit ergibt sich, wenn die ^{15}N-Einbaumengen relativiert werden. Summiert man die ^{15}N-Mengen der einzelnen Aminosäuren im Tierkörper unter Eliminierung der jeweiligen Traceraminosäure (Gruppe 1: Methionin; Gruppe 2: Lysin; Gruppe 3: Glycin + Serin) so können die prozentualen Anteile der einzelnen Aminosäuren am gesamten transaminierten ^{15}N berechnet werden. Die entsprechenden Werte sind in Tab. 6 zusammengestellt. Die Werte stellen ein relatives Maß für die Intensität dar, mit der die einzelnen Aminosäuren ^{15}N durch Transaminierung aufnehmen. Zwischen den vier Gruppen ergibt sich praktisch bei allen Aminosäuren eine recht gute Übereinstimmung. Das läßt den Schluß zu, daß unabhängig von der Art der Tracerverbindung und unabhängig von der Art des Futters stets die gleiche Rangfolge in der Beteiligung der Aminosäuren an der Transaminierung besteht. Demnach sind Glutaminsäure, Arginin[1]), Glycin, Asparaginsäure und Alanin am stärksten, Threonin, Phenylalanin, Histidin und Lysin am geringsten beteiligt. Diese Rangfolge wurde bereits 1946 von *Schoenheimer* [53] nach Verabfolgung von ^{15}N-Leucin an Ratten gefunden und stimmt auch mit anderen Literaturangaben [2, 20, 30, 47, 48] überein.

4. Schlußfolgerungen für die Eignung der ^{15}N-Verbindungen als Tracer bzw. als Markierungssubstanz

Bei Einsatz von (^{15}N)-Lysin verbleiben 90 ... 95% des inkorporierten ^{15}N in Form von Aminosäuren-^{15}N, davon

[1]) Bei Arginin muß jedoch, wie angeführt, der Großteil der ^{15}N-Markierung auf direkte Umwandlungsprozesse zurückgeführt werden.

Tab. 6. Prozentualer Anteil der einzelnen Aminosäuren an der Summe des Amino-^{15}N im Tierkörper (unter Eliminierung der ^{15}N-Menge der jeweiligen Tracer-Aminosäure)

Aminosäuren	Gruppe 1 DL-(^{15}N)-Methionin		Gruppe 2 L-(^{15}N)-Lysin		Gruppe 3 (^{15}N)-Glycin[1])		Gruppe 4 $(^{15}NH_4)_2SO_4$		Mittelwert $\bar{x} \pm s$
	Tier 2	Tier 3	Tier 4	Tier 5	Tier 7	Tier 9	Tier 10	Tier 11	
Glu	17,2	18,5	14,5	17,0	21,1	19,8	20,3	21,9	18,8 ± 2,5
Arg	(17,3)[2])	(16,7)[2])	(12,0)[2])	(17,1)[2])	(16,5)[2])	(16,2)[2])	(19,6)[2])	(20,0)[2])	17,0 ± 2,4[2])
Ala	10,9	10,5	11,9	10,8	14,2	14,9	10,7	9,6	11,7 ± 1,9
Gly	9,5	12,5	12,6	7,7	–	–	10,9	7,7	10,2 ± 2,2
Asp	11,2	9,9	11,1	8,7	9,1	9,4	7,9	8,9	9,5 ± 1,2
Leu	7,9	7,8	6,9	7,0	10,5	11,0	6,6	7,0	8,1 ± 1,7
Ser	5,4	4,6	5,9	4,4	–	–	4,7	4,1	4,9 ± 0,7
Val	5,3	5,1	5,3	6,2	4,0	4,7	4,3	4,5	4,9 ± 0,7
Pro	3,1	2,3	1,8	4,5	5,1	4,3	4,5	4,5	3,8 ± 1,2
Ile	3,2	3,3	3,9	3,6	2,8	3,3	2,8	3,4	3,3 ± 0,4
His	2,5	2,1	5,8	6,0	1,5	1,6	1,5	1,6	2,8 ± 1,9
Lys	1,8	2,5	–	–	6,7	6,9	2,3	2,7	3,8 ± 2,3
Zys	2,4	2,3	1,2	0,5	(5,9)[2])	(4,7)[2])	0,7	0,7	2,3 ± 2,0
Met	–	–	0,6	2,8	1,6	1,1	1,3	1,4	1,6 ± 0,7
Phe	1,9	1,5	(6,0)	2,3	0,7	0,8	0,8	1,1	1,3 ± 0,6
Thr	0,5	0,5	0,5	1,5	0,4	1,1	0,9	1,0	0,8 ± 0,4
Σ	100		100		100		100		
Davon in ess. AS	42,7	41,7	42,3	46,9	50,6	51,6	41,0	43,3	45,0 ± 4,2
in nicht-ess. AS	57,3	58,3	57,7	53,1	49,4	48,4	59,0	56,7	55,0 ± 4,2

[1]) Eliminierung von Gly + Ser
[2]) ^{15}N-Markierung ist nicht oder nur teilweise durch Transaminierung bedingt.

der weitaus größte Anteil unverändert als Lysin-^{15}N. Markiertes Lysin repräsentiert wegen des Fehlens von Nebenreaktionen in fast idealer Weise den Lysinstoffwechsel. Bei Kenntnis des Lysingehaltes im Körper kann daraus die Proteinsyntheserate des Gesamtkörperproteins ermittelt werden. Die Oxidationsrate und die Reutilisationsrate für den Gesamtkörperstickstoff sind dagegen mit denen des Lysins nicht identisch.

Analoges gilt für Threonin, das ebenfalls kaum an Transaminierungs- und anderen Umwandlungsprozessen teilnimmt (vgl. Tab. 6).

^{15}N-markierte Aminosäuren, die ihren ^{15}N-Stickstoff durch Transaminierung rasch und zu einem unterschiedlich großen Anteil auf andere Aminosäuren übertragen, sind zur Charakterisierung des Gesamtkörperstickstoffs bedingt geeignet. Da in diesem Falle jedoch niemals alle Aminosäuren homogen markiert werden, stellt jede Gabe einer einzelnen ^{15}N-markierten Aminosäure als Tracer immer einen Kompromiß zwischen der Markierung der Aminosäure und einer homogenen Markierung des Proteins dar und die so bestimmten Stoffwechselparameter haben deshalb stets nur Näherungscharakter.

Nachteilig wirkt bei der Bestimmung von Proteinstoffwechselparametern durch ^{15}N-Tracer-Gabe auch der oft relativ hohe Anteil an ^{15}N, der über Nebenreaktionen in andere Verbindungen eingebaut wird. Beide Einschränkungen bestehen auch für den Einsatz von (^{15}N)-Glycin als Traceraminosäure [52]. Daß mit (^{15}N)-Glycin ermittelte Proteinstoffwechselparameter trotzdem oft mit den auf anderem Wege bestimmten Daten gut übereinstimmen [11, 13, 38, 49] scheint daran zu liegen, daß die durch eine höhere Harnausscheidungsrate für Glycin-Stickstoff bedingte Unterschätzung der Syntheserate durch die Einbeziehung von Glycin in Nichtprotein-Stoffwechselwege annähernd kompensiert wird [2]. (^{15}N)-Glycin hat gegenüber anderen ^{15}N-Traceraminosäuren den Vorteil einer hohen Verwertungsrate, einer raschen Übertragung des markierten Stickstoffs auf andere Aminosäuren [45] und einen ökonomischen Vorteil [1]. Da aber nicht a priori mit einem Ausgleich der genannten Fehlermöglichkeiten gerechnet werden kann, ist der Einsatz von (^{15}N)-Glycin allenfalls zur Bestimmung relativer Stoffwechselverhältnisse zu akzeptieren.

Bei Einsatz einer einzelnen ^{15}N-Traceraminosäure zur Bestimmung von Gesamtkörperstoffwechseldaten über das 3-Pool-Modell von *Sprinson* und *Rittenberg* stellt (^{15}N)-Glycin offenbar noch den besten Kompromiß dar, da keine andere Aminosäure allein bessere Stoffwechseldaten liefern kann. Diese Einschätzung zeigt die Möglichkeiten aber auch die Grenzen des Einsatzes von (^{15}N)-Glycin als Tracer.

Auf Grund des hohen Transaminierungsgrades und der raschen Übertragbarkeit des ^{15}N auf andere Aminosäuren würden noch (^{15}N)-Glutaminsäure und (^{15}N)-Alanin als Traceraminosäuren in Frage kommen. Sie sind aber wesentlich teurer und die für (^{15}N)-Glycin geschilderten Nachteile bestehen — sogar in verstärktem Maße — auch für diese Aminosäuren.

Die Verwendung von DL-(^{15}N)-Methionin als Tracer kann wegen des Umbaues von markiertem D-Methionin zu unmarkiertem L-Methionin und wegen der relativ hohen Transaminierungsrate weder für die Untersuchung des Proteinstoffwechsels noch für die Bestimmung des Methioninstoffwechsels empfohlen werden. Die Verwendung von L-(^{15}N)-Methionin als Tracer wird sicherlich günstigere Ergebnisse liefern, was aber in weiteren Versuchen noch geprüft werden muß.

Zur Bestimmung genauerer Werte und für die Ermittlung absoluter Proteinstoffwechselraten empfiehlt sich der Einsatz eines Gemisches von ^{15}N-markierten Aminosäuren [36, 38, 50], eines ^{15}N-markierten Proteins [12, 36] oder eines Hydrolysates von ^{15}N-markiertem Protein [51]. Der Vorteil liegt darin, daß hier die gleiche ^{15}N-Dosis von vornherein auf ein breites Spektrum von ca. 20 Aminosäuren verteilt ist, während sie mit Gabe einer einzelnen markierten Aminosäure nur als deren α-Amino-N verabreicht wird [50].

^{15}N-Ammoniumsalze stellen für die Markierung eines Organismus mit ^{15}N gegenüber ^{15}N-markierten Futtermitteln eine günstige Alternative dar. Die prozentuale Verteilung des Amino-^{15}N aus $^{15}NH_4$-Sulfat betrug in unserem Versuch 42% für die essentiellen Aminosäuren und 58% für die nichtessentiellen Aminosäuren. Damit empfiehlt sich die Methode von *Simon* u. a. [17] und *Bergner* u. a. [19, 20] zur Markierung von Organismen mit ^{15}N-markierten Ammoniumsalzen sowohl aus ökonomischen Gründen als auch wegen der relativ gleichmäßigen ^{15}N-Verteilung auf essentielle und nichtessentielle Aminosäuren. In Bezug auf die Verteilung des inkorporierten ^{15}N auf essentielle und nichtessentielle Aminosäuren bestehen zwischen den 4 Varianten keine wesentlichen Unterschiede, wenn die Traceraminosäure und die aus ihr durch direkte chemische Umwandlung gebildeten Aminosäuren ausgeschlossen werden. Das Verhältnis entspricht in etwa der Verteilung von unmarkierten essentiellen und nichtessentiellen Aminosäuren im Tierkörper. In Abhängigkeit von der gewählten Traceraminosäure verbleibt jedoch ein unterschiedlich großer Anteil der Markierung in der inkorporierten Traceraminosäure, beim (^{15}N)-Methionin 12%, bei (^{15}N)-Glycin 32% und bei (^{15}N)-Lysin 84%.

Eingegangen am 22. 10. 1985

Literatur

[1] *H. Faust, H.-D. Czarnetzki, W. Hartig,* ZfI-Mitt. **71** (1983) 49

[2] *J. Waterlow, P. J. Garlick, D. C. Millward,* Proteinturnover in mammalian tissues and in the whole body. Amsterdam — New York — Oxford: North-Holl. Publ. Comp. (1978)

[3] *P. J. Garlick, D. J. Millward, W. P. T. James,* Biochem. J. **136** (1973) 935

[4] *O. Simon, R. Münchmeyer, H. Bergner,* Br. J. Nutr. **40** (1978) 243

[5] *M. Florkin, A. Neuberger, L. L. M. van Deenen,* Comprehensive Biochemistry. 19 B, Part 1: Protein Metabolism, Amsterdam — Oxford — New York: Elsevier Scientific Publishing Comp. (1980)

[6] *P. J. Garlick,* In: *Florkin, Neuberger, van Deenen,* Protein Metabolism (1980), 77

[7] *M. J. Rennie, D. Halliday,* Proc Nutrit. Soc. **43** (1984) 189

[8] *D. Halliday,* Advances in the measurement of stable isotopes In: Nitrogen Metabolism in Man. *J. C. Waterlow* u. *J. M. L. Stephen,* Applied science publishers, London, New Jersey 1981, 295

[9] *W. Hartig, H. D. Czarnetzki, H. Faust, K. Hirschberg, K. Jung, H. Junghans, K. Wetzel, H. Willgerodt,* Fortschritte der Medizin, DDR-Med.-Rep. **7** (1978) 563

[10] *K. Krawielitzki, T. Völker, S. Smulikowska, H.-D. Bock, J. Wünsche,* Arch. Tierernähr. **27** (1977) 609

[11] *M. H. N. Golden, A. A. Jackson,* in: *Waterlow* and *Stephen*: Nitrogen Metabolism in Man. London, New Jersey, Appl. Sci. Publ. 1981, 323

[12] *W. Heine, J. Richter, C. Plath, K. Wutzke, H.-J. Stolpe, M. Thiess, J. Töwe,* Fortschritte der Medizin, DDR-Med.-Rep. **12** (1983) 10, 838

[13] *D. Picou, T. Taylor-Roberts, J..C. Waterlow,* J. Physiol. **200** (1969) 52

[14] *R. Köhler, T. Zebrowska, G. Gebhardt,* Arch. Tierernähr. **28** (1978) 317

[15] *H. Faust, K. Jung, P. Krumbiegel,* Isotopenpraxis **17** (1980) 193

[16] *M. Hernandez, O. Simon, H. Bergner,* Arch. Tierernähr. **31** (1981) 651
[17] *O. Simon, P. Bohley, H. Bergner, R. Münchmeyer,* Arch. Tierernähr. **34** (1984) 341
[18] *R. Schiemann, H.-D. Bock, J. Keller, L. Hoffmann, K. Krawielitzki, M. Klein,* Arch. Tierernähr. **33** (1983) 57
[19] *U. Bergner, H. Bergner, O. Simon,* Arch. Tierernähr. **34** (1984) 505
[20] *H. Bergner, U. Bergner, V. Adam,* Arch. Tierernähr. **34** (1984) 353
[21] *J. Voigt, B. Piatkowski, Z. Ceresnakova, R. Krawielitzki, K. Adam,* Arch. Tierernähr. **34** (1984) 387
[22] *J. Voigt, B. Piatkowski, R. Krawielitzki, A. Sommer, Z. Ceresnakova, H. Engelmann, J. Vancisin, H. Girschewski, L. Chrastinova,* Arch. Tierernähr. **33** (1983) 327
[23] *E. Kolb,* Arch. Tierernähr. **13** (1963) 264
[24] *K. Wetzel, G. Hübner, O. Gebhardt,* Arch. Tierernähr. **13** (1963) 424
[25] *H. Faust, H. Gürtler, G. Hübner, H. Hübner, H. Mielke, W. Rommel, M. Ulbrich, K. Wetzel,* Arch. Tierernähr. **13** (1963) 475
[26] *A. A. Jackson, M. H. N. Golden,* Clin. Sci. **58** (1980) 517
[27] *H. Wu, C. W. Bishop,* J. Appl. Physiol. **14** (1959) 1
[28] *D. P. Tschudy, H. Bacchus, S. Weissmann, D. M. Watkin, M. Eubanks, J. White,* J. Clin. Invest. **38** (1959) 892
[29] *M. Taruvinga, A. A. Jackson, M. H. N. Golden,* Clin. Sci. **57** (1979) 281
[30] *T. Völker, K. Krawielitzki, J. S. Keller, M. Klein,* Arch. Tierernähr. **33** (1983) 189
[31] *K. Krawielitzki, T. Völker,* Isotopenpraxis **14** (1978) 152
[32] *R. Köhler, G. Gebhardt,* Arch. Tierernähr. **17** (1967) 485
[33] *W. B. Souffrant, R. Köhler, G. Gebhardt,* Arch. Tierernähr. **31** (1981) 35
[34] *J. Wünsche, H.-D. Bock, U. Hennig, F. Kreienbring, E. Borgmann,* Arch. Tierernähr. **29** (1979) 221
[35] *B. Schumann,* Untersuchungen zur idealen Stickstoff- und Aminosäurenresorption sowie zur endogenen Stickstoffsekretion aus dem Pankreas beim Schwein. Diss. A Karl-Marx-Univ. Leipzig, 1985
[36] *K. Krawielitzki,* Das Kompartementmodell und seine Bedeutung für Untersuchungen über den ernährungsphysiologischen Proteinstoffwechsel bei Nichtwiederkäuern. Diss. B, Akademie der Landwirtschaftswiss. 1984
[37] *K. Krawielitzki, R. Schadereit, T. Zebrowska, J. Wünsche, H.-D. Bock,* Arch. Tierernähr. **34** (1984) 589
[38] *K. Krawielitzki,* Tag.-Ber., Akad. Landwirtschaftswiss. DDR (1983) 217, 51
[39] *H. Ebersdobler,* in: *Scheunert* und *Trautmann,* Lehrbuch der Veterinär-Physiologie, 6. Aufl. Berlin: Parey-Verlag 1976, 241
[40] *J. Leibetseder,* Degussa-Informationsdienst für die Mischfutterindustrie **1** (1984) 1
[41] *P. L. Pellet, H. Kaba,* J. Nutr. **102** (1972) 61
[42] *R. Köhler, H. Müller, J. Hentzel, G. Gebhardt,* Wiss. Z. KMU Leipzig, Math.-Nat. Wiss. R **27** (1973) 277
[43] *H.-D. Bock,* Arch. Tierernähr. **33** (1983) 520
[44] *T. P. Stein,* in: *Waterlow* and *Stephen.* Nitrogen Metabolism in Man. London, New Jersey, Appl. Sci. Publ. (1981) 345
[45] *J. M. Conway, D. E. Matthews, V. R. Young, D. M. Bier,* Americ. J. Clin. Nutr. **33** (1980) 919
[46] *E. Hofmann,* Intermediärstoffwechsel, Dynamische Biochemie, Teil III, 2. Aufl. Berlin: Akademie Verlag 1971
[47] *D. E. Matthews, J. M. Conway, V. R. Young, D. M. Bier,* Metabolism. Clin. Exp. **30** (1981) 886
[48] *K. Gruhn, A. Hennig, H. Graf,* Arch. Tierernähr. **34** (1984) 539
[49] *M. H. N. Golden, J. C. Waterlow,* Clin. Sci. Mol. Med. **53** (1977) 277
[50] *W. Heine, J. Richter, Chr. Plath, K. Wutzke, F. Oswald, S. Oswald, J. Töwe,* J. Pediatr. Gastroenter. u. Nutrit. **2** (1983) 599
[51] *K. Wutzke, W. Heine,* Isotopenpraxis **21** (1985) 99
[52] *A. Neuberger,* in: *Waterlow* und *Stephen,* Nitrogen Metabolism in Man. London, New Jersey, Appl. Sci. Publ. (1981) 253
[53] *R. Schoenheimer,* The Dynamic State of Body Constituents. Havard Univ. Press, Cambridge, Mass. 1946

Isotopenpraxis **22** (1986) 8, S. 269 bis 272

Messungen spezifischer Tritiumaktivitäten von Luftfeuchte und Niederschlag — Methodik und Anwendung

R. Lewis[1]), *K. Fröhlich*[2]), *D. Hebert*[2])

[1]) Zentralinstitut für Kernforschung der Akademie der Wissenschaften der DDR, Rossendorf bei Dresden
[2]) Bergakademie Freiberg, Sektion Physik

Die Ergebnisse von Tritiummessungen an Luftfeuchte und Niederschlägen von Freiberg der Jahre 1982/83 werden vorgestellt und Probenahmemethodik sowie -apparaturen beschrieben.
Der Verlauf der Zeitreihen von Luftfeuchte- und Niederschlagsdaten sowie deren Beeinflussung durch technogene Tritiumquellen werden diskutiert und die Nutzung der Analysendaten für Untersuchungen atmosphärischer Feuchtetransportprozesse aufgezeigt.

The results of tritium measurements of precipitation and air humidity samples collected in Freiberg, G.D.R., during the time period from 1982 to 1983 are presented and the sampling technique is described.
The seasonal variation of the measured specific tritium activities and the influence of technogenic tritium are discussed. The application of this tritium data to investigations of atmospheric humidity transport processes is pointed out.

Keywords

air samplers; atmospheric precipitations; earth atmosphere; humidity; sample preparation; sampling; tritium

1. Einleitung

Seit der Entwicklung der Tritiummethode zur Datierung von unterirdischen Wässern [1] auf der Grundlage der Messung der Tritiumverhältnisse[3]) von Niederschlägen wurde wiederholt auf die Bedeutung und Notwendigkeit von Tritiumanalysen atmosphärischen Wasserdampfes hingewiesen [2, 3, 4].

Der z. T. hohe apparative Aufwand, welcher für eine Luftfeuchtebeprobung erforderlich ist, verhinderte bisher eine breite Anwendung dieser Verfahrensweise.

Die Bestimmung von Tritium (als HTO) in Niederschlag und Luftfeuchte ermöglicht eine kontinuierliche Gewinnung von Daten zum regionalen und lokalen atmosphärischen Tritiumniveau sowie dessen temporären Variationen.

Anschriften: [1]) DDR-8051 Dresden PF 19, [2]) Bernhard-von-Cotta-Straße 4, DDR-9200 Freiberg

[3]) Tritiumverhältnis (engl. Abk. T. R. — tritium ratio); T.R. = 1 bedeutet 1 ^{3}H-Kern/10^{18} H-Kerne und entspricht einer spezifischen Tritiumaktivität von 0,118 $Bqkg^{-1}$.

Dieses Datenmaterial ist Voraussetzung für Untersuchungen troposphärischer HTO-Verteilungsmechanismen, welche neben anderen eine Grundlage für die komplexe Erforschung von Feuchtetransportprozessen und der Zirkulation in der Atmosphäre bilden.

2. Probenahme und Messung

Seit 1969 werden die Tritiumverhältnisse der Monatsniederschläge von Freiberg bestimmt [5, 6].

Ab Mitte Februar 1982 wurde mit der Beprobung der Luftfeuchte begonnen. Die Extraktion des Wasserdampfes aus der Luft erfolgte durch Ausfrieren. Zunächst wurde eine einfache stationäre, später eine transportable, automatische (Abb. 1) Probenahmeapparatur entwickelt und eingesetzt [7].

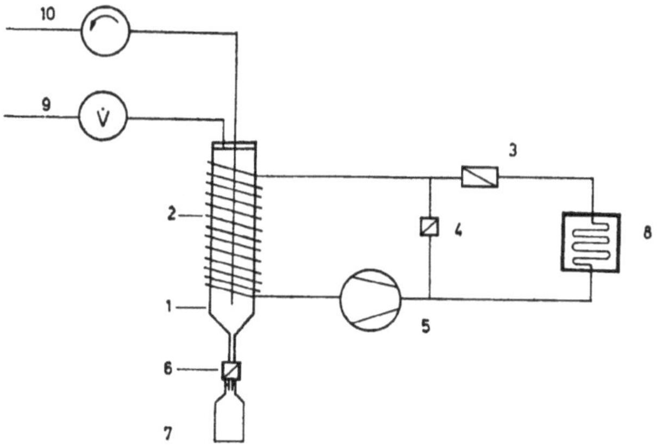

Abb. 1. Schematischer Aufbau der automatischen Apparatur zur Luftfeuchteextraktion. 1 – Kühlfalle aus Kupfer; 2 – Verdampfer; 3 – thermisches Ventil zur Kühltemperaturoptimierung; 4 – Magnetventil zur Steuerung des Kühl- und Abtauvorganges; 5 – Verdichter; 6 – Magnetventil zur Entleerung der Kühlfalle nach dem Abtauen; 7 – Sammelbehälter; 8 – Verflüssiger; 9 – Gaszähler; 10 – Pumpe

Im ersten Fall wurde die Luft durch eine mit einem Aceton-Trockeneis-Gemisch gekühlte gläserne Kühlfalle geleitet, im zweiten ein handelsüblicher Verdichtersatz zur Kühlung verwendet. Die aus Kupfer gefertigte Kühlfalle wurde zu diesem Zweck mit mehreren Windungen des Verdampferrohres (Abb. 1) umgeben. Kühl- und Abtauvorgang sind wahlweise per Hand oder automatisch steuerbar. Die erreichten Kühltemperaturen betrugen −80 bzw. −50 °C bei einem maximal zulässigen Luftdurchsatz ($\dot{V}$) von 0,12 bzw. 0,16 $m^3\,h^{-1}$. Während ein- bis dreitägiger Feuchtesammelperioden ($\dot{V} = 0{,}06\ m^3\,h^{-1}$) waren entsprechend der absoluten Luftfeuchtigkeit 5 bis 30 ml Wasser gewinnbar. Zeitlich parallel dazu wurde der gefallene Niederschlag aufgefangen.

Die Probenahmeperioden von 1 bis 3 Tagen waren so gewählt worden, daß eine hinreichend gute Auflösung der HTO-Variationen in Abhängigkeit von Änderungen der meteorologischen Verhältnisse zu erwarten war.

Für Probenahmezeiten < 1 h wurde ein adsorptives Verfahren angewendet. Hierzu wurde Molekularsieb der freien Atmosphäre ausgesetzt. Von 75 g regeneriertem Molekularsieb konnten innerhalb dieses Zeitraumes ca. 5–6 g Wasser adsorbiert werden. Anschließend erfolgte die Rückgewinnung des Wassers mit einer Vakuumdesorptionsapparatur [7].

Die Tritiummessungen wurden nach einer speziellen Probenvorbereitung [8] mit dem Proportionalzählrohr durchgeführt. Der geringe Probenumfang gestattete keine elektrolytische Anreicherung von Tritium für die Messungen. Die Nachweisgrenze des Meßverfahrens lag bei T. R. = 5.

3. Ergebnisse und Diskussion

Die für Mitte Februar bis August 1982 vorliegenden Tritiumverhältnisse der Luftfeuchte (C_v) sind in Abb. 2 dargestellt. Zum Vergleich wurden in diese Darstellung die Monatsmittel der Tritiumverhältnisse des Niederschlages ($\overline{C}_p$) aufgenommen.

Innerhalb der einzelnen Monate wiesen die Luftfeuchtedaten eine beträchtliche Variabilität, gekoppelt mit Änderungen der Wetterlage, auf. Im Februar, März und Juli lagen die Tritiumwerte der Luftfeuchte über den jeweiligen Monatsmitteln der Niederschläge. Für März und Juli konnte dieser Sachverhalt auf die geringe Anzahl von Niederschlagsereignissen zurückgeführt werden, wobei die Niederschläge des Monatsanfangs am ergiebigsten und durch niedrige Tritiumverhältnisse gekennzeichnet waren. In der zweiten Februarhälfte fiel kein Niederschlag, jedoch nur für diese wurden die Tritiumverhältnisse der Luftfeuchte bestimmt.

Demnach beschreiben die Monatsmittel der Niederschläge das allgemeine Tritiumniveau nur unvollständig, die entsprechenden Luftfeuchtedaten hingegen geben dieses lückenlos wieder. Eine Übereinstimmung zwischen beiden unter Berücksichtigung der Isotopenfraktionierung ist umso wahrscheinlicher, desto gleichmäßiger verteilt Niederschläge über den zu betrachtenden Zeitraum fallen. Im Jahresdurchschnitt ist diese Bedingung für humides Klima erfüllt. Für detaillierte Untersuchungen kurzzeitiger (Stunden, Tage) atmosphärischer Prozesse mittels Tritium, wie sie Feuchtetransport und Zirkulation darstellen, ist dies eine unzulässige Näherung.

Der durch Bombentritium bedingte jahreszeitliche Gang der Tritiumverhältnisse atmosphärischer Feuchte [9] mit einem Minimum im Winter und Maximum im Frühjahr konnte anhand der Luftfeuchtedaten nicht festgestellt werden, da die bisher vorliegenden Meßergebnisse nur den Zeitraum Februar bis August umfassen. Monatsmittel der Tritiumverhältnisse der Luftfeuchte für 1981/82 der Station Heidelberg (BRD) [4] geben den saisonalen Trend ebenfalls nur sehr schwach wieder. Ursache dessen ist dort seine Überlagerung durch technogenes Tritium lokalen und regionalen Ursprungs. Dieser Fakt wird durch das mehrfache, kurzzeitige Auftreten erhöhter Tritiumverhältnisse der Luftfeuchte bestätigt. Die Monatsmittel der Tritiumverhältnisse des Niederschlages von Freiberg und Heidelberg widerspiegeln den Trend deutlicher, da sie weit weniger durch lokale Tritiumquellen beeinflußt werden.

Tritiumverhältnisse von T. R. < 20 traten immer in Verbindung mit straffen Westwindlagen auf. Sie sind typisch für das HTO-Niveau ozeanischer Feuchte und ermöglichen eine Indikation selbiger. Demgegenüber waren die Tritiumverhältnisse osteuropäischer Feuchte meist durch hohe Werte (T. R. = 40–50) charakterisiert.

Den Nachweis der Beeinflussung der Freiberger Feuchtedaten durch lokale Quellen erbrachte der Vergleich der Tritiumverhältnisse parallel gesammelter Luftfeuchte- und Niederschlagsproben. Im $C_v - C_p$-Diagramm (Abb. 3)

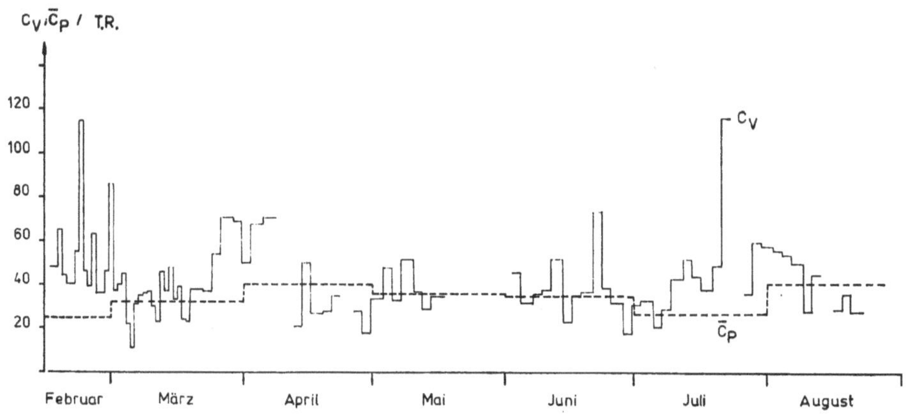

Abb. 2. Tritiumverhältnisse der Luftfeuchte (C_V; 14. 2.–19. 3. 82 = Tages-; 19. 3.–25. 8. 82 = Zwei- und Dreitagesmittelwerte) und des Niederschlages ($\bar{C}_P$; Monatsmittel) von Freiberg

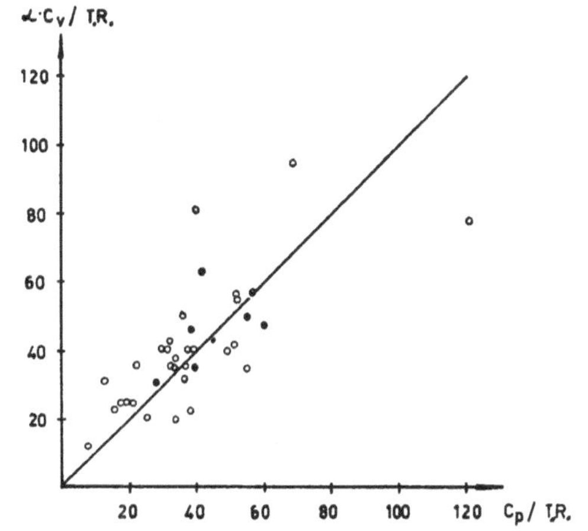

Abb. 3. $C_V - C_P$-Diagramm von Freiberg (○ 1982; ● 1983; Probenahmeperiode jeweils 1 bis 3 Tage)

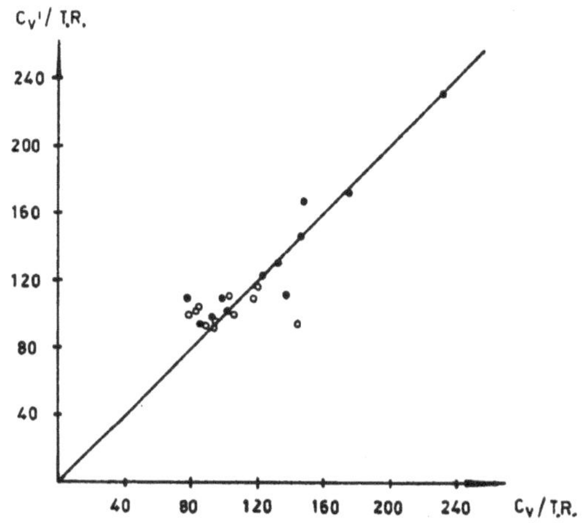

Abb. 4. Gemessene (C_V) und berechnete (C_V') Tritiumverhältnisse in der Umgebung der Tritiumquelle. ○ – bei Normalbetrieb; ● – bei Targetwechsel

indizieren weit über der Gleichgewichtsgeraden ($C_p = \alpha \times C_v$, $\alpha = 1{,}1$) liegende Wertepaare lokales technogenes Tritium. Als Quelle konnte ein in einer Entfernung von 850 m zur Probenahmestelle betriebener Neutronengenerator ermittelt werden.

Durch zwei Feldexperimente, zum einen während des Normalbetriebes, zum anderen während eines Targetwechsels, wurde die HTO-Verteilung am Boden in der Umgebung der Quelle ermittelt. Die Probenahme erfolgte durch Adsorption des Wasserdampfes an Molekularsieb über ca. eine Stunde.

Bei bekannter Quellenstärke Q [$\mathrm{Bqs^{-1}}$] und stationären Bedingungen können die Tritiumverhältnisse (C_v') der bodennahen Feuchte in der Nähe der Quelle vorausberechnet werden [10].

$$C_v' = \frac{Q}{f \cdot \pi \cdot v \cdot \sigma_y(x) \cdot \sigma_z(x) \cdot k} \times \exp\left[-\frac{1}{2}\left(\frac{y^2}{\sigma_y^2(x)} + \frac{h^2}{\sigma_z^2(x)}\right)\right] \qquad (1)$$

f – absolute Luftfeuchtigkeit in $\mathrm{kgm^{-3}}$, v – mittlere Windgeschwindigkeit in $\mathrm{ms^{-1}}$, h – Höhe des Emissionspunktes in m, x; y – Koordinaten in bzw. senkrecht zur Ausbreitungsrichtung in m; σ_y; σ_z – horizontaler bzw. vertikaler Dispersionskoeffizient in m; $k = 0{,}118\ \mathrm{Bqkg^{-1}}$.

Für $x \leq 10^3$ m wurden die Dispersionskoeffizienten für Stadtgebiete gemäß den Diffusionskategorien nach [11] und aus [12] erhalten. Die Korrelation von gemessener (C_v) und berechneter (C_v')-HTO-Verteilung erbrachte eine gute Übereinstimmung (Abb. 4).

Die vorliegenden Ergebnisse sowie die Auswertung der Windverhältnisse für den Zeitraum Februar bis August gestatteten die Ausgliederung von Meßwerten aus der Langzeitreihe der Tritiumverhältnisse der Luftfeuchte, welche durch lokales industrielles HTO beeinflußt waren (Tab. 1). Insgesamt belief sich deren Anteil auf weniger als 10%. Infolgedessen konnten die bereinigten Analysedaten für weiterführende Untersuchungen von regionalen raumzeitlichen HTO-Variationen und damit von Feuchtetransportprozessen genutzt werden.

Die Zuordnung der Tritiumverhältnisse der Freiberger Luftfeuchte zu den jeweiligen Luftmassen, denen sie entstammte (Abb. 5), ergab ihre Abhängigkeit von den Luftmassentrajektorien und deren Länge über dem Festland, wobei die Transportgeschwindigkeit der Feuchte zu berücksichtigen war.

Maritime Feuchte mit kurzer kontinentaler Transportstrecke der nördlichen Luftmassen mP_A und mP wies niedrige Tritiumverhältnisse auf. Kontinentale Feuchte (Luftmassen: cT, cP_T, cP, cP_A), welche weite Strecken über das Festland geführt wurde, hatte nahezu einheitlich hohe Tritiumwerte. Eine starke Variabilität der Tritiumverhältnisse wurde für die Feuchte westeuropäischer Luft-

Tab. 1. Durch eine lokale Tritiumquelle erhöhte Tritiumverhältnisse in der Luftfeuchte von Freiberg

lfd. Nr.	T	C_{vg}/T.R.	C_{Pg}/T. R.	ΔC/T. R.
1	×	65		17
2		63		26
3		86	69	23
4		46	36	13
5	×	48		17
6	×	74	40	38
7		57	42	19

T = Targetwechsel am Neutronengenerator; C_{vg}, C_{Pg} = gemessene Tritiumverhältnisse von Luftfeuchte und Niederschlag (Probenahmeperiode 1 bis 3 Tage); ΔC aus $C_v = C_{Pg}/1.1$ und $C_{vg} - C_v = \Delta C$ bzw. anhand anderer Luftfeuchtemeßergebnisse berechnete Abweichung vom Normalwert (C_N), $C_N = C_{vg} - \Delta C$

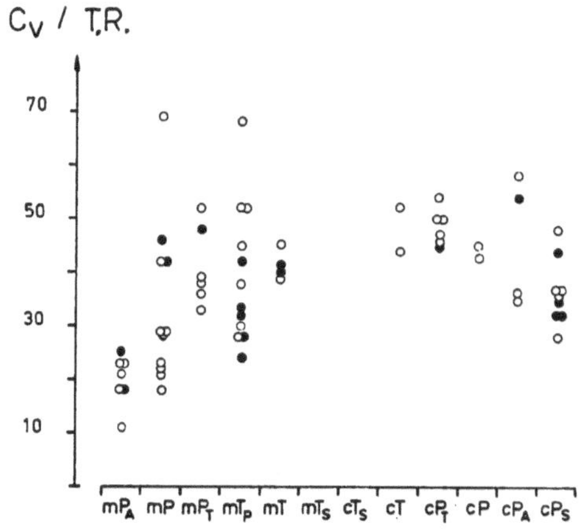

Abb. 5. Tritiumverhältnisse des Wasserdampfes (C_V) verschiedener europäischer Luftmassen im Raum Freiberg (○ 1982, ● 1983; Probenahmeperiode jeweils 1 bis 3 Tage)

m_PT = Erwärmte-; mP = Grönländische-; mP_A = Arktische-; cP_S = Skandinavische-; cP_A = Nordsibirische-; cP = Festlands-; cP_T = Rückkehrende Polarluft; cT = Kontinentale-; cT_S = Afrikanische-; mT_S = Mittelmeer-; mT = Atlantische-; mT_P = Meeres-Tropikluft

massen (mT_P, mP_T, mT, mP) festgestellt. Hierfür war die Freisetzung von regional wirksamem technogenem Tritium über Westeuropa verantwortlich.

Weitere Untersuchungen in der angegebenen Richtung sollten eine quantitative Auswertung der Analysedaten ermöglichen.

Eingegangen am 6. 11. 1985

Literatur

[1] *S. Kaufman, W. F. Libby*, The natural distribution of tritium. The Physical Review **93** (1954) 1337

[2] *K. Kigoshi, K. Yoneda*, Daily variations in the tritium concentration of atmospheric moisture, J. Geophys. Res. **75** (1970) 2981

[3] *H. G. Östlund, A. S. Mason*, Atmospheric distribution of HTO and HT. Proc. of the international conference on low-radioactivity measurements and applications, held at High Tatras, Czechoslovakia, 6th to 10th October 1975, Bratislava 1977, S. 252

[4] *H. Schoch-Fischer* et al., Hydrometeorological factors controlling the time variation of D, ^{18}O and ^{3}H in atmospheric water vapour and precipitation in the northern westwind belt. In: Isotope Hydrology 1983, IAEA, Wien 1984, p. 3 – 30

[5] Technical reports series No. 226, Environmental isotope data No. 7, World survey of isotope concentration in precipitation 1976 – 1979, Wien: IAEA 1983

[6] Technical reports series, Environmental isotope data No. 8, World survey of isotope concentration in precipitation, Wien: IAEA. (to be published)

[7] *R. Lewis*, Beiträge zur Untersuchung und Nutzung von atmosphärischem Tritium, unveröffentlichte Diss., Bergakademie Freiberg 1985

[8] *K. Fröhlich* u. a., Radioaktive Umweltisotope in der Hydrogeologie, Freiberger Forschungsh., R.C, Leipzig 1977

[9] *W. Weiss* et al., Evidence of pulsed discharges of tritium from nuclear energy installations in central European precipitation, Proc. of a Symp. on Behaviour of Tritium in the environment, held at San Francisco 16. – 20. 10. 1978 Wien: IAEA 1979

[10] *K. I. Vogt*, Dispersion of airborn radioactivity released from nuclear installations population exposure in local and regional environment, Proc. of a Symp. on Physical Behaviour of radioactive Contaminants in the Atmosphere, held at Vienna. Nov., 12 – 16, 1973, Wien: IAEA 1974

[11] *F. A. Gifford*, Use of routine meteorological observations for estimating atmospheric dispersion. Nuclear Safety **2** (1961) 47

[12] *R. P. Hosker*, Estimates of dry deposition and plume depletion over forests and grassland. Proc. of a Symp. on Physical Behaviour of radioactive Contaminants in the Atmosphere, held at Vienna Nov., 12 – 16, 1973, Wien: IAEA 1974

Isotopenpraxis 22 (1986) 8, S. 272

Buchbesprechung

Lasertechnik im Bauwesen

V. A. Kougija, Berlin: VEB Verlag für Bauwesen. 1984; 143 S., 98 Abb., 3 Tab.; kartoniert M 18.—

Ausgehend von der Darstellung des Laserprinzips und optischer Systeme für die Laserstrahlbündelung werden für geodätische Messungen im Bauwesen eingesetzte laserbestückte Geräte wie Laserfluchtgeräte, Rotationslaser, Lasertheodolite u. a. und ihre Prüfung beschrieben. Nach allgemeinen geodätischen Anwendungen von Lasergeräten wie Übertragung einer Bezugslinie mit vorgegebener Längsneigung werden spezielle Anwendungen im Brückenbau (einschließlich Deformationsmessungen), im Tunnelbau sowie Absteckungsarbeiten bei der Vorfertigung von Segmenten für Unterwassertunnel und der Ausrichtung von Grundbaggern im Hafenbau zur Kenntnis gebracht. Ein Abschnitt über die Anwendung geodätischer Lasergeräte im Wohnungs- und Industriebau, z.B. zur Vertikalität von Turmbauten, als Meßsystem für den Kühlturmbau, zur Überwachung der Montage von Antennenträgern, Fassadenelementen schließt die Beschreibung praktischer Einsatzfälle der Lasertechnik im Bauwesen mit ein.

Prinzipskizzen und fotografische Abbildungen einzelner Geräte, die Angabe von Typen- und Leistungsdaten aus dem internationalen Angebot und konkrete Erfahrungen beim praktischen Einsatz vermitteln anschaulich, was auf diesem Gebiet heute Standard ist. Die mathematische Beschreibung von Zusammenhängen ist in der Darlegung extrem kurz und in den Grundrechenarten gehalten. Das Literaturverzeichnis orientiert auf erreichbare deutschsprachige Quellen.

Die Monographie ist als eine Einführung in die geodätische Lasermeßtechnik im Bauwesen angelegt, die neben zahlreichen technischen Gerätedetails hilft, konkret Einsatzbedingungen, Anwendungsmöglichkeiten und Leistungsgrenzen dieser Technik einzuschätzen. *H. Baumbach*

Isotopenpraxis 22 (1986) 8, S. 273 bis 275

Untersuchungen über die Auslösung von Spannungsrißkorrosion durch Adsorption von Chloriden an hochlegiertem Cr-Ni-Stahl

G. Hartmann, C. Jonas, P. Kulicke, A. Walter

(Akademie der Wissenschaften der DDR, Zentralinstitut für Isotopen- und Strahlenforschung, Leipzig)[1])

Es wird ein Verfahren zur Bestimmung der Adsorption von Chloridionen an unterschiedlich vorbehandelten CrNi-Stahloberflächen aus wäßrigen NaCl-Lösungen im Konzentrationsbereich bis herab zu 0,8 ppm unter Verwendung des Radionuklids ^{36}Cl beschrieben. Die örtliche Verteilung des Chlorids auf der Oberfläche wird durch autoradiographische Abbildung ermittelt. Die Experimente wurden im Temperaturbereich um 250 °C durchgeführt.
Die Abhängigkeit der Chloridabsorption von der Chloridkonzentration wird dargestellt und der Zusammenhang zwischen adsorbierter Chloridmenge und -verteilung, mechanischer Spannung und Korrosionsanriß diskutiert.

A method for determination of chloride ions adsorbed on the surface of stainless steel from aqueous solutions of NaCl at concentrations down to 0,8 ppm using the radionuclide ^{36}Cl is described. The local distribution of the chloride ions on the surface is determined by autoradiographic imaging. The experiments are performed in a temperature range about 260 °C in an autoclave.
The dependency of chloride-adsorption on the preparation of the surface and chloride concentration is shown and the connection between adsorbed amount and distribution of chloride, mechanical tension and the beginning of corrosion is discussed.

Keywords

adsorption; autoclaves; autoradiography; chlorine ions; chlorine-36; density; distributions; images; magnesium chlorides; sodium chlorides; stainless steel; stress corrosion; surfaces

Einleitung

Die Spannungsrißkorrosion hängt im wesentlichen von Werkstoffzusammensetzung und -vorbehandlung, Größe und zeitlichem Verlauf der vorhandenen bzw. aufgeprägten Spannungen im Material sowie Temperatur, Druck und chemischer Zusammensetzung des umgebenden Mediums, insbesondere der Cl^-- und O_2-Konzentration, ab [1].

Der Mechanismus der Auslösung der Spannungsrißkorrosion ist noch nicht vollständig aufgeklärt. Das bis heute vorliegende und außerordentlich umfangreiche Datenmaterial zahlreicher systematischer Untersuchungen ist dafür oft nur bedingt aussagefähig. Die Ergebnisse sind infolge einer Vielzahl unterschiedlicher Versuchsparameter und Materialien nicht oder nur schwierig vergleichbar, zu verallgemeinern oder auf einen speziellen Fall zu übertragen.

Der unzureichende Erkenntnisstand ist u. E. auch darauf zurückzuführen, daß Spannungsrißkorrosion in örtlich sehr eng begrenzten Gebieten der entsprechenden Werkstoffoberflächen ihren Ursprung hat, in denen die Mediumbedingungen und auch der Oberflächenzustand im Verlauf des Prozesses nur schwer zu erfassen sind und von integral gemessenen Werten erheblich abweichen können.

Nachweisverfahren

Ziel der Arbeiten war die hochempfindliche und örtlich hochauflösende Bestimmung von Cl^- auf Edelstahloberflächen mit Hilfe einer Radionuklidtechnik. Das Versuchsprinzip ist, alle Untersuchungen zur Cl^--Adsorption mit dem Radionuklid ^{36}Cl durchzuführen und die $^{36}Cl^-$-Verteilungen auf den Stahloberflächen mit Hilfe der Autoradiographie zu erfassen [2]. Die Leistungsfähigkeit dieser Versuchsmethodik läßt sich wie folgt abschätzen:

Für die Schwärzung S in einem Autoradiogramm gilt Gleichung (1):

$$S = E \cdot A_S \cdot t \cdot F \quad (1)$$

mit E = Filmempfindlichkeit

A_S = spezifische Aktivität des verfügbaren Radionuklids

t = Expositionszeit

F = Flächenmasse des Chlors auf der Stahloberfläche

Für die Nachweisgrenze des Verfahrens für Chlor ergibt sich damit Gleichung (2):

$$F_{Grenz} = \frac{S_{min}}{E A_S t_{max}} \quad (2)$$

mit $S_{min} = 0{,}01$

$t_{max} = 10^7$ s

$A_S = 10^8$ Bq g^{-1} (Chlor)

$E = 1{,}4 \cdot 10^{-8}$ cm^2 s^{-1} Bq^{-1}

wird $F_{Grenz} \approx 10^{-9}$ g (Chlor) cm^{-2}, d. h. eine monomolekulare Chloridbelegung ist sicher erfaßbar.

Ein weiterer Vorteil des autoradiographischen Nachweises beruht auf der Tatsache, daß die Schwärzung von der Flächenaktivität, d. h. der Chloridflächenmasse abhängt, nicht aber von der Größe der Fläche auf der die Chloridverteilung abgebildet wird.

[1]) Anschrift: DDR-7050 Leipzig, Permoserstraße 15

Die Angabe 10^{-9} g (Chlor) cm^{-2} bedeutet z. B. auch 10^{-11} g (Chlor) mm^{-2} oder 10^{-15} g (Chlor) $(10\ \mu m)^{-2}$.

Um Aussagen über die örtliche Chloridverteilung auf den Edelstahloberflächen im Bereich der Gefügestruktur machen zu können, sind hochauflösende autoradiographische Abbildungen der entsprechenden $^{36}Cl^-$-Verteilungen nötig. Die Herstellung solcher Mikroautoradiogramme erfordert einen möglichst engen Kontakt zwischen Probe und dem autoradiographischen Material, der entweder durch Aufgießen der flüssigen Photoemulsion direkt auf die Probe oder durch Aufbringen trägerfreier Emulsionsschichten unter Wasser realisiert werden kann. Bei diesen „Naßverfahren" kommt es sehr leicht zu chemischen Reaktionen zwischen der Metalloberfläche und der gegenüber chemischen Einflüssen extrem empfindlichen Photoemulsion, die beim Entwicklungsprozeß zu Schwärzungen führen und damit $^{36}Cl^-$ vortäuschen. Es macht sich deshalb erforderlich, den direkten Kontakt zwischen Probe und Photoemulsion auszuschließen, ohne dabei die Abbildungsschärfe zu beeinträchtigen. Deshalb wurde eine Technik entwickelt, mit der eine geeignete Schutzschicht auf die Probenoberfläche aufgebracht werden kann, die so dünn ist, daß einerseits die Abbildungsunschärfe nicht sehr steigt und andererseits chemische Wechselwirkungen mit der Photoemulsion ausgeschlossen werden.

Damit besteht die Voraussetzung, hochauflösende autoradiographische Abbildungen von $^{36}Cl^-$-Verteilungen auf den Probenoberflächen herzustellen, die Aufschluß über eine Korrelation zwischen $^{36}Cl^-$-Adsorption und Gefügestruktur der Proben geben sollen

Die quantitative Erfassung der $^{36}Cl^-$-Verteilungen auf den Probenoberflächen setzt die quantitative Auswertung der Autoradiogramme voraus [3]. Um der Schwärzung im Autoradiogramm die entsprechende Chloridflächenmasse mit einem möglichst geringen Fehler zuordnen zu können, wurden Eichpräparate hergestellt, die stets gemeinsam mit den Proben exponiert werden, so daß der Einfluß der Fehlerquellen Filmempfindlichkeit und Entwicklungsbedingung eliminiert ist. Die Eichpräparate bestehen aus dünnen Plastfolien, in die definierte Menge $n^{36}Cl^-$ homogen eingebracht wurden. Diese Folien sind auf eine Stahlplatte aufgezogen, so daß auch bezüglich der rückgestreuten Strahlung die gleichen Verhältnisse vorliegen, wie bei den zu untersuchenden Proben.

Damit wurde die Voraussetzung geschaffen, bei weiteren Untersuchungen die Chloridflächenmasse auf den Probenoberflächen im ng-Bereich mit einem Fehler von ca. 8% zu bestimmen.

Versuchsdurchführung und Ergebnisse

Zur Vereinheitlichung der Proben im Hinblick auf Reproduzierbarkeit und Vergleichbarkeit der Ergebnisse erfolgte eine Probenvorbehandlung durch Lösungsglühen, Schleifen oder Elektropolieren und Passivieren.

Neben spannungsfreien Proben wurden auch solche untersucht, auf die eine Biegespannung bis zur Fließgrenze oder auch mittels einer speziellen Vorrichtung definierte Zugspannungen bis zu 550 Nmm^{-2} aufgebracht wurden. Nach Versuchsende wurden die Proben einem normierten Wasch- und Spülprozeß mit Wasser, Aceton und Ethanol unterzogen, um zu vermeiden, daß anhaftende Reste des Elektrolyten auf der Oberfläche verbleiben und adsorbiertes $^{36}Cl^-$ vortäuschen.

In Vorversuchen wurde bei Normaldruck und Temperaturen um 100 °C in wäßrigen $MgCl_2$-Lösungen mit Einwirkungszeiten bis zu 120 h und bei wäßrigen $NaCl_2$-Lösungen bis zu 500 h gearbeitet. Die Chloridkonzentrationen betrugen bei $MgCl_2$ 12, 19 und 26 Gew.%. Die spezifische Aktivität betrug 6,7 MBq g^{-1} Chlor.

Mit der geschilderten Versuchstechnik und unter den genannten Versuchsbedingungen ließen sich $^{36}Cl^-$-Verteilungen auf den Probenoberflächen erzeugen, und ihre örtliche Verteilung konnte autoradiographisch bestimmt werden. In Abhängigkeit von Temperatur, Konzentration und mechanischer Spannung wurden in $MgCl_2$-Lösungen die Zeiten bis zum Auftreten von Korrosionseffekten (Risse, Lochfraß) ermittelt (Abb. 1 und Abb. 2). In NaCl-Lösungen trat unter diesen Bedingungen keine erkennbare Korrosion auf.

Die Auswertung der Autoradiogramme ergab bei den mit NaCl-Lösungen behandelten Proben auch nach längeren Versuchszeiten eine weitgehend homogene $^{36}Cl^-$-Verteilung über die gesamte Probenoberfläche; die Chloridflächenmasse lag hier bei nur $2 \cdot 10^{-8}$ g cm^{-2} (Abb. 3).

Bei den mit $MgCl_2$-Lösungen behandelten Proben ist das Chlorid bevorzugt an den Stellen lokaler Korrosion zu finden; die Chloridflächenmasse liegt bei $2 \cdot 10^{-4}$ g cm^{-2} (Abb. 4). Die lokalen Korrosionseffekte traten hier bevor-

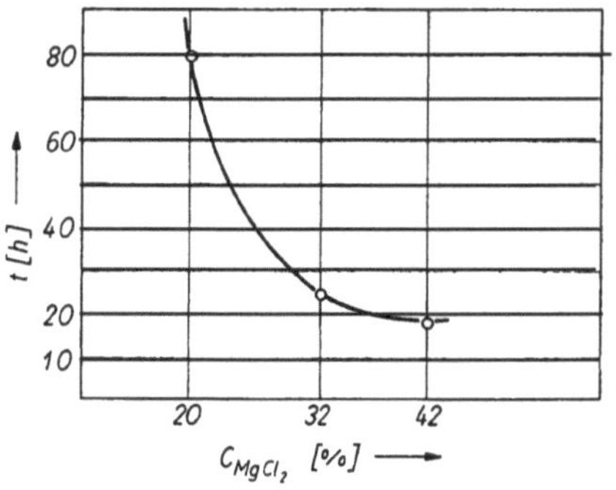

Abb. 1. Zeit bis zum Auftreten von korrosiven Anrissen in Abhängigkeit von der $MgCl_2$-Konzentration, System: $Mg^{36}Cl_2$, X8CrNiTi 18.9, mechanische Spannung $\sigma = 200\ N \cdot mm^{-2}$

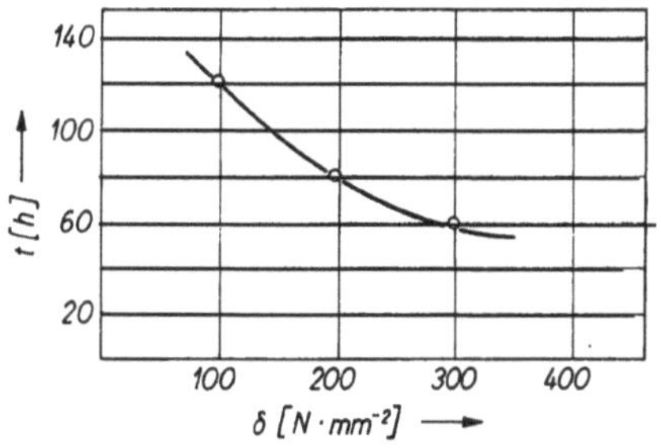

Abb. 2. Zeit bis zum Auftreten von korrosiven Anrissen in Abhängigkeit von der Biegespannung, System: $Mg^{36}Cl_2$, X8CrNiTi 18.9, Temperatur $T = 378$ K

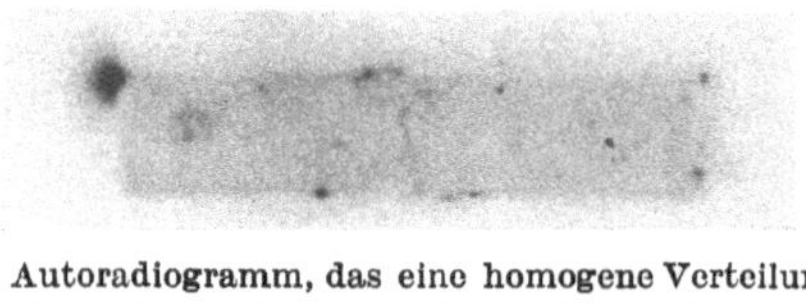

Abb. 3. Autoradiogramm, das eine homogene Verteilung von $^{36}Cl^-$ nach Einwirkung einer 6,8%igen NaCl-Lösung bei $T = 373$ K zeigt

Abb. 4. Adsorbiertes $^{36}Cl^-$ an Lochfraßstellen (Autoradiogramm)

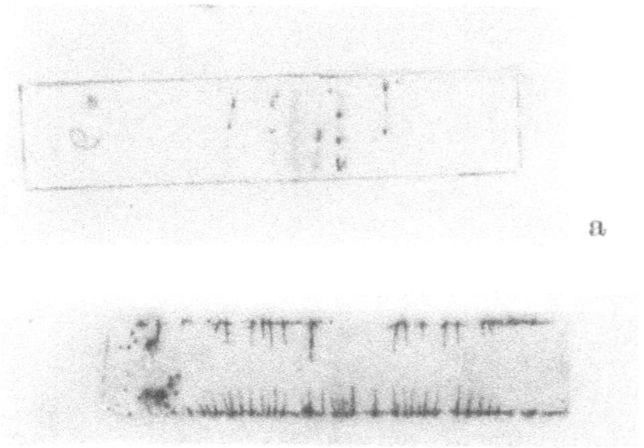

a

b

Abb. 5. Adsorbiertes $^{36}Cl^-$ an Rissen und Lochfraßstellen (Autoradiogramme) a) im Gebiet größter Biegespannung b) in Gebieten größter Scherspannungen

zugt in Gebieten erhöhter Biegespannung oder Scherspannung (Abb. 5) auf.

Zur Realisierung eines Temperaturbereiches bis nahe 300 °C wurde in einem Autoklaven gearbeitet. Um in vertretbaren Zeiten Effekte erzielen und untersuchen zu können, wurden Versuche mit der maximal möglichen spezifischen Aktivität des verfügbaren $Na^{36}Cl$ von 6,7 MBq g^{-1} Chlor durchgeführt. Die Konzentration dieses kommerziell erhältlichen $Na^{36}Cl$ betrug 84 mg/ml. Den Abmessungen des für ein Volumen von 0,8 cm^3 gebauten Mikroautoklaven angepaßt, wurden Proben von 3 mm Breite und 0,5 mm Dicke mit Biegespannung beaufschlagt. Die Werte der Biegespannungen reichten bis zur Fließgrenze.

Die $^{36}Cl^-$-Einwirkung erfolgte im Elektrolyten bei sich frei einstellendem elektrochemischem Potential. Der Sauerstoffgehalt im Elektrolyten während des Versuches lag entsprechend der Löslichkeit bei Werten $< 0{,}1$ ppm.

Bei den Versuchen wurden Einwirkungszeiten des Elektrolyten von 72 h bis 600 h realisiert. Der Konzentrationsbereich der eingesetzten Elektrolyte erstreckte sich von 8,4% bis 0,84 ppm Chlorid. Die Probenvorbehandlung erfolgte wie bei den Vorversuchen beschrieben; gleiches gilt für den Spülprozeß zu Versuchsende.

Die Proben wiesen Chloridbelegungen auf der gesamten Oberfläche auf. Dem Konzentrationsbereich von 8,4% bis 0,84 ppm Chlorid im Elektrolyten entsprechen Chloridflächenmassen auf den Probenoberflächen von $2 \cdot 10^{-5}$ g cm^{-2} bis $4{,}2 \cdot 10^{-7}$ g cm^{-2}. Das bedeutet, daß bei hohen Chloridkonzentrationen des Elektrolyten die Chloridadsorption der Probenoberfläche unter den gewählten Versuchsbedingungen einem Sättigungswert zustrebt. An Stellen lokaler Korrosion tritt eine verstärkte Chloridadsorption auf (Abb. 6).

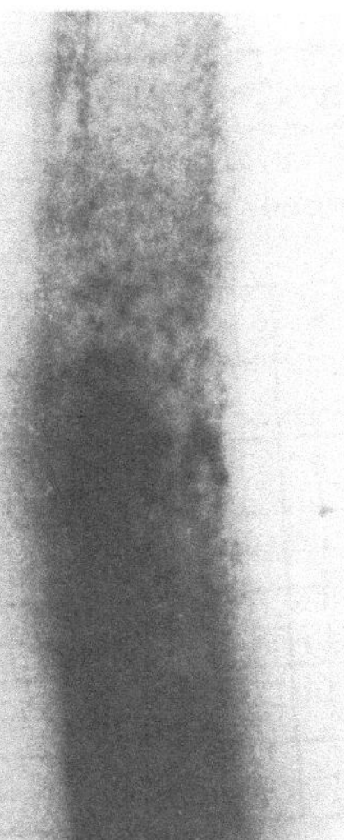

Abb. 6. Autoradiogramm mit einer Chloridverteilung auf der Oberfläche der Probe und Chloridaufkonzentrationen an Stellen größter Biegespannung

Im Mittel wurde für das Verhältnis der mittleren Chloridflächenmassen bei Spannungen an der Fließgrenze zu denen in ungespannten Gebieten ein Wert von 3,2 ermittelt.

Die Chloridadsorption in Gebieten erhöhter Spannung ist inhomogen. Die mikroskopische Abbildung der Autoradiogramme zeigt deutlich runde, etwa 30 µm bis 300 µm große Zentren erhöhter Chloridkonzentration (Abb. 6).

Die lichtoptische Abbildung und das Autoradiogramm sind bezüglich dieser Zentren nicht korrelierbar. Die Empfindlichkeit der entwickelten Methode reicht aus, um auch in Wasser mit Chloridgehalten im Konzentrationsbereich um 1 ppm Cl^-, d. h. unter praxisnahen Bedingungen, die adsorbierte Chloridmenge und ihre örtliche Verteilung auf der Metalloberfläche zu bestimmen.

Weiterführung der Arbeiten

Mit dem erprobten Verfahren soll in weiteren Arbeiten der Einfluß der Sauerstoffkonzentration sowie die Effektivität von Korrosionsinhibitoren hinsichtlich der Adsorption von Chlorid untersucht werden. Außerdem soll die Empfindlichkeit des Nachweisverfahrens durch Einsatz von $Na^{36}Cl$ mit einer höheren spez. Aktivität um eine Größenordnung gesteigert werden, so daß die bisher relativ langen Expositionszeiten bis zu mehreren Wochen entsprechend verkürzt werden können. Vorgesehen ist auch der Einsatz eines Autoklaven, in dem Untersuchungen zur Adsorption von Chlorid bei einem praxisnahen Wärmefluß in Spalten möglich sind, um Chloridanreicherungseffekte in solchen Gebieten zu erfassen.

Weiterhin wurden Arbeiten zur Modifizierung des Nachweisverfahrens begonnen mit dem Ziel, nichtradioaktive Chloridverteilungen zu erfassen. Das kann z.B. durch nachträgliche Markierung des Chlorids durch Isotopenaustausch mit radioaktivem $Na^{36}Cl$ erfolgen. Wird dabei eine Lösung genügend hoher Aktivität angeboten, dann wird die auf der Oberfläche vorhandene Chloridmenge im Autoradiogramm sicher erfaßt.

Um Radionuklide mit für die autoradiographische Abbildung geeigneten Parametern für den Isotopenaustausch zu verwenden, wird gegenwärtig der Einsatz des neuerdings vom ZfI hergestellten kurzlebigen Radionuklids ^{39}Cl ($t_{1/2}$ = 56,2 min), das mit sehr viel höherer spezifischer Aktivität hergestellt werden kann, diskutiert. Für die Untersuchungen mit Isotopenaustausch käme aber auch das Radionuklid ^{82}Br in Frage, das aber nur ähnliche Eigenschaften wie ^{36}Cl hat und deshalb, was die Adsorption an der Oberfläche des Materials anbelangt, andere Informationen geben könnte. Über entsprechende Untersuchungen wird an anderer Stelle berichtet werden.

Eingegangen am 14. 11. 1985

Literatur

[1] *B. M. Gordon,* Materials Performance **19** (1980) 29
[2] *R. Reinhardt,* Isotopenpraxis **18** (1982) 41
[3] *W. Hermann, G. Hartmann, H. Baumbach,* Atompraxis **14** (1968) 1

Isotopenpraxis 22 (1986) 8, S. 276 bis 279

Labelling of Di-iodofluorescein with Iodine-131 by Ultraviolet Excitation

M. Raieh, E. M. Mikhail

(Nuclear Chemistry Department, Nuclear Research Centre, Atomic Energy Establishment, Cairo)[1]

The labelling of diiodofluorescein with ^{131}I using the ultraviolet waves as an excitation energy has been tried. Different normal alcohols have been used as solvents. The different parameters affecting the labelling yield such as, presence of water in different proportions, pH of the medium and presence of different concentrations of carrier, have been studied. Also, the effective lines of the Hg lamp have been elucidated. The labelling mechanism in different conditions has been explained. The study revealed that higher yields were obtained when using a mixture of ethanol and water as solvent at pH $\sim$ 5.

Es wurde eine Markierung von Di-iodofluorescein mit ^{131}I unter Einstrahlung von UV-Licht als Anregungsenergie versucht. Verschiedene normale Alkohole wurden als Lösungsmittel verwendet. Unterschiedliche Parameter, welche die Markierungsausbeute beeinflußten, wie die Gegenwart von Wasser in unterschiedlichen Verhältnissen, der pH-Wert und die Anwesenheit von unterschiedlichen Trägerkonzentrationen, sind untersucht worden. Ferner sind die effektiven Linien einer Hg-Lampe ermittelt worden. Der Markierungsmechanismus unter den verschiedenen Versuchsbedingungen kann erklärt werden. Die Untersuchung hat ergeben, daß höhere Ausbeuten dann erhalten werden, wenn eine Mischung von Ethanol und Wasser als Lösungsmittel bei einem pH-Wert von ungefähr 5 eingesetzt wird.

Keywords

excitation; fluorescein; iodine compounds; iodine 131; labelled compounds; labelling; ultraviolet radiation

1. Introduction

Diiodofluorescein labelled with radioiodine-131 ($DI^{131}F$), was considered as an organic radiopharmaceutical due to its application as a diagnostic agent in the field of nuclear medicine. It was usually used for the detection of brain and ovarian tumours [1–4]. *Taplin* et al. [5] were the first that introduced the iodine labelled diiodofluorescein ($DI^{131}F$) into clinical medicine in 1955 as a mean of testing liver function.

The different techniques so far applied for the labelling of DIF are based on either synthesis, or exchange reactions, however, the main labelling technique for such compound is by isotopic exchange [6]. In spite of its low solubility in aqueous media trials have been done to label it in different buffer solutions at pH range from 5–6·5 [7–8]. Organic media such as carbon tetrachloride [9] and ethanol [10] have been tried.

In this study a trial has been done to label DIF using the ultraviolet excitation technique in normal alcohols and in a mixture of water and alcohols media for the exchange reaction.

2. Experimental

2.1. Materials

Diiodofluorescein was obtained from B.D.H. Carrier free radioiodine-131 was produced locally and supplied in the form of NaI^{131}. Different solutions of NaI^{131} in different alcohols have been prepared [11].

All the other reagents used in this study are of analytical grade.

2.2. Equipments

a) High pressure mercury lamp (Gallenkamp ultraviolet discharge lamp H-530) and the following filters:
   1. Filter 6543: Nickel oxide 9863 Broad spectrum, 3 mm thick, non heat resistant, absorbs visible and transmits 0% 2300 Å, 45% 2500 Å, 90% 3000 Å and 0% 4200 Å.
   2. Filter 16223 B: (Red purple 5874). Heat resistant, 5 mm thick, absorbs visible and transmits 0% 3000 Å, 55% 3600 Å and 0% 4100 Å.
   3. Filter 6544: Very dense 5860 narrow spectrum, 5 mm thick, absorbs maximum visible and transmits 0% 3100 Å, 38% 3600 Å, 0% 3990 Å.

b) Gamma counter type ST 6.

c) Spectrophotometer Perkin Elmer 330.

2.3. Labelling Technique

50 mg DIF were dissolved in 5 ml absolute ethanol (or other solvent) were introduced into a quartz tube which is placed at 3 cm distance from the ultraviolet lamp. The temperature in the tube was kept constant (30 ± 2 °C) by continuous ventilation. To the tube is introduced about 0·1 mCi NaI^{131} in ethnaol (or other solvent) and the tube was stoppered tightly. Samples were withdrawn at different time intervals considering the moment of addition NaI^{131} as the zero time. The samples are subjected to descending paper chromatography to estimate the labelling yield at each time. The solvent used for chromatography was ethanol:ammonia (1:1):water in the ratio 25:5:70 (8, 9). On the starting line of the chromatographic paper (Whatmann 3 MM) was primarily placed a spot of Na_2SO_3 + $Na_2S_2O_3$ dil. solution to reduce any oxidation products of radioiodine present in the samples to be studied.

[1]) Address: Cairo, Egypt, P.O. Box 13759

Kurznachrichten /

Newlets

Lebensmittelbestrahlung:

BUNDESRAT ZUR LEBENSMITTELBESTRAHLUNG

Mit einer kleinen Anfrage hatte sich Nationalraetin Monika Weber (LdU, Zuerich) erkundigt, wie sich die Landesregierung zur Strahlenkonservierung von Lebensmitteln stelle und ob die Bestrahlung nicht unerwuenschte Nebenwirkungen verursache. Am 16.12.1985 nahm der Bundesrat zu diesem Vorstoss wie folgt Stellung:

"Die Behandlung der Lebensmittel mit ionisierenden Strahlen ist ein technologisches Verfahren zur verbesserten Haltbarkeit von Lebensmitteln. Die Forschungsanstrengungen zur Lebensmittelbestrahlung haben die urspruenglich hohen Erwartungen bisher nur teilweise erfuellt. Internationale Experten-Kommissionen (FAO/WHO/IAEA) haben sich wiederholt mit moeglichen Gesundheitsrisiken bestrahlter Lebensmittel auseinandergesetzt. 1980 wurde festgehalten, dass - unter Einhaltung gewisser Bedingungen bei der Bestrahlung - derart behandelte Nahrungsmittel als fuer den Menschen bekoemmlich betrachtet werden koennen; die radioaktive Bestrahlung hat keine Radioaktivitaet des Lebensmittel zur Folge. Mit ionisierenden Strahlen behandelte Lebensmittel duerfen nach der Lebensmittelverordnung erst in den Verkehr gebracht werden, wenn das Verfahren vom Bundesamt fuer Gesundheitswesen als zulaessig anerkannt ist. Die konventionellen Methoden zur Verbesserung der Halbarkeit der Lebensmittel sind heute so hoch entwickelt, dass dem Konsumenten auch ohne das neue Verfahren der Bestrahlung hochwertige und sichere Lebensmittel angeboten werden koennen. Der Bundesrat vertritt daher die Auffassung, dass derzeit keine schwerwiegenden, zwingenden Gruende vorliegen, welche eine allgemeine Zulassung der Lebensmittelbestrahlung rechtfertigen wuerden. Die Behandlung bestimmter Lebensmittelgruppen (z.B. Gewuerze) mit ionisierenden Strahlen bedarf in jedem Einzelfall der Pruefung, ob eine solche Behandlung notwendig ist. Derzeit ist trotz grosser Anstrengungen noch keine Nachweismethode bekannt, mit welcher sich routinemaessig eindeutig nachweisen laesst, ob ein Lebensmittel bestrahlt worden ist." (SVA-Bulletin (1986)2)

JAPAN - KEINE BEDENKEN ZUR STRAHLENKONSERVIERUNG GEEIGNETER FISCHPRODUKTE UND WURSTSORTEN

Die Strahlenbehandlung von verarbeitetem Fischfleisch und gewissen Wurstsorten ist gesundheitlich unbedenklich und konserviert die genannten Lebensmittel, ohne deren Geschmack zu beeintraechtigen. Zu dieser Schlussfolgerung gelangte in Japan eine spezielle Untersuchungskommission in ihrem Bericht zuhanden der Atomic Energy Commission (AEC). Interressierte Verarbeiter haben nunmehr die Moeglichkeit, dem japanischen Ministerium fuer Gesundheit und Wohlfart entsprechende Projekte zur Genehmigung zu unterbreiten. Zurzeit wird die Strahlenkonservierung in Japan in industriellem Ausmass nur bei Kartoffeln angewandt. Die dafuer errichtete Bestrahlungsanlage in Shihoro auf der Insel Hokkaido mit Co-60 als Strahlenquelle koennte nach Auffassung von Fachleuten auch fuer die Bestrahlung von Fischprodukten und Wuersten eingesetzt werden. Nach Angaben der Atomic Energy of Canada Ltd. ist die Strahlenkonservierung von tiefgefrorenem Fisch und von Fischprodukten bereits in Holland, Chile und Bangladesh erlaubt. (SVA-Bulletin (1986) 3)

USA - FDA-BEWILLIGUNG FUER BESTRAHLUNG VON OBST UND GEMUESE

Die amerikanische Lebensmittelbehoerde, die Food and Drug Administration (FDA), hat kuerzlich die Bewilligung erteilt, frisch geerntetes Obst und Gemuese zu bestrahlen, um solcherart den Schaedlingsbefall zu bekaempfen und die Haltbarkeit der Produkte zu verbessern. Margaret M. Heckler, bisherige Direktorin des US-Department of Health and Human Services, betonte in diesem Zusammenhang, die Strahlenanwendung koenne zu einer Verringerung des Einsatzes von Pestiziden beitragen und gewisse Lebensmittel ueberdies verbilligen. Mrs. Heckler fuegte bei, dass die Bestrahlung im Unterschied zu den Pestiziden keine Rueckstaende in der Nahrung hinterlasse und fuer die Verbraucher ohne Risiko sei. Nach den Vorschriften der FDA muessen strahlenbehandelte Lebensmittel als solche deklariert sein. Die Anwendung ionisierender Strahlen wurde in den USA erstmals vor rund zwei Jahrzehnten zur Vernichtung von Schaedlingen im Weizen und zur Verlangsamung des Auskeimens der Kartoffeln bewilligt. Kraeuter und Gewuerze duerfen seit 1983, Schweinefleisch seit 1985 bestrahlt werden. (SVA-Bulletin (1986)4)

Hochtemperaturreaktor:

THTR-300 AM NETZ

Als weltweit erstes grosstechnisches Prototyp-Kernkraftwerk mit Kugelhaufenreaktor hat die Anlage THTR-300 am 16.11.1985 erstmals Strom in das oeffentliche Netz eingespeist. Damit ist die Inbetriebnahme des Hochtemperaturreaktors in Hamm-Uentrop, BRD, in ihre entscheidende Phase getreten. Abgeschlossen werden soll sie Anfang 1986 mit der offiziellen Uebergabe der Anlage an die Betreiberin, die Hochtemperatur-Kernkraftwerk GmbH (HKG). Beim ersten Hochfahren erreichte das Kernkraftwerk eine Leistung von 80 MW. Die Faehigkeit, zusaetzlich zur Stromerzeugung nukleare Prozesswaerme auf hohem Temperaturniveau fuer eine Vielzahl von industriellen Prozessen zur Verfuegung zu stellen, macht Hochtemperaturreaktoren fuer das Ruhrgebiet und aehnliche Standorte wirtschaftlich interessant, vor allem auf dem Gebiet der Kohleveredlung. Als Nachfolgereaktor befindet sich gegenwaertig der HTR-500 in der Planung, an der erstmals auch potentielle Anwender von HTR-Verfahrenstechniken beteiligt sind. (SVA-Bulletin (1985) 21)

Strahlentechnik:

KONSERVIERUNG VORGESCHICHTLICHER HOLZFUNDE

Die Arbeit "Strahlenchemische Konservierung und Auswertung von Holz aus der Wallanlage von Stillfried" von A. Kies, P. Reithmayr und R. Schaudy von Institut fuer Ur- und Fruehgeschichte der Universitaet Wien erschien im Dezember 1985 als Bericht OEFZS-4340 des Oesterreichischen Forschungszentrum Seibersdorf. Sie beschreibt die Konservierung von Holzresten der urnenfelderzeitlichen Wallanlage von Stillfried a.d. March, Niederoesterreich, durch Impraegnierung mit Methylmethacrylat und darauf folgende Haertung mit Gammastrahlung sowie die Anfertigung von Duennschliffen aus den verfestigten Holzstuecken. Die Arbeit zeigt, dass das Verfahren der strahlenchemischen Konservierung zur Bewaeltigung bisher unueberwundener Schwierigkeiten bei der Praeparation von Holzfunden in problematischem Erhaltungszustand taugt. Die Strahlenkonservierung ermoeglichte die Bestimmung der Holzarten bzw. -gattungen im Holzrahmenwerk des Stillfrieder Westwalls aus der Periode Hallstatt B. Hauptsaechlich identifiziert wurden Eichen, Linde und Ulme, was Rueckschluesse auf die Zusammensetzung des Eichenmischwaldes der genannten Periode erlaubt. (SVA-Bulletin (1986)5)

Strahlenschutz:

GROSSBRITANNIEN: HAEUFUNG VON ZWISCHENFAELLEN IN SELLAFIELD

Innerhalb von weniger als vier Wochen, in der Zeit vom 23.1. bis 18.2.1986 haben sich in der Anlage zur Wiederaufarbeitung verbrauchter metallischer Brennelemente der British Nuclear Fuels plc. (BNFL) in Sellafield drei Stoerfaelle ereignet, bei denen radioaktive Stoffe freigesetzt wurden. Am 23.1. war aufgrund fehlerhafter Bedienung gereinigtes Uran in die See abgeleitet worden, am 5.2. erfolgte eine Freisetzung von Pu ueber die Raum- und Abluft unmittelbar nach der Wartung einer Probenahmestation und

und am 18.2. fuehrte ein Rohrbruch zum Austritt von schwachradioaktivem Lagerbecken-Wasser in einen Rohrkanal. Wegen einer moeglichen Strahlenbelastung von Betriebspersonal und Umgebung konzentrierte sich das Interesse der Aufsichtsbehoerde, des Nuclear Installations Inspectorate (NII), auf das Ausmass und die Folgen der Pu-Freisetzung am 5.2. In einer ersten offiziellen Stellungnahme des NII vom 14.2. wird der Stoerfall als geringfuegig hinsichtlich seiner Wirkung auf die Umgebung eingestuft. Nach den bisherigen Untersuchungen koenne angenommen werden, dass der Stoerfall ausserhalb des Gelaendes der Anlage keine Auswirkungen gehabt habe. Die Frage, wie viele der 71 Operateure, die am 5.2. in der Anlage taetig waren, Pu inhaliert haben und welches Ausmass moegliche Kontaminationen haben, wird fruehestens in zwei Monaten, die fuer die laufende Ueberwachung der Betroffenen erforderlich sind, exakt beantwortet werden koennen. Hatte BNFL zunaechst nur von zwei Arbeitern gesprochen, bei denen der Verdacht einer Kontamination bestuende, so sind es jetzt 15 der Operateure, bei denen dieser Verdacht besteht. Aufgrund erster Messungen ist ein Arbeiter bei dem Stoerfall mit der maximal zulaessigen Jahresdosis belastet worden. Weniger die wohl geringfuegigen Auswirkungen der Stoerfaelle fuer die Sicherheit, als deren Haeufigkeit haben in weiten Kreisen Grossbritanniens, aber auch des benachbarten Nordirland zu Beunruhigung gefuehrt. Am 19.2. hat der Health and Safety Executive als oberste britische Gesundheits-Behoerde eine eingehende Ueberpruefung von Sellafield angeordnet. Die Ueberpruefung wird von 12 Inspektoren des NII durchgefuehrt werden und wird auf mindestens sechs Monate Dauer geschaetzt. (SVA-Bulletin (1986)5)

KEINE STRAHLENGEFAHR AM BILDSCHIRMARBEITSPLATZ

Aufgrund fundierter Untersuchungen hat die deutsche Gesellschaft fuer Strahlen- und Umweltforschung Muenchen (GSF) nochmals bekraeftigt, dass an Arbeitsplaetzen vor Bildschirm- und Datensichtgeraeten keine Strahlengefaehrdung besteht und dass auch TV-Apparate kein Strahlenrisiko darstellen. Wie die GSF nachwies, verhindert das dicke Glas der Bildroehre im Unterschied zu Roentgenroehren jegliches Austreten der im Innern als Stoerstrahlung entstehenden Roentgenstrahlen. Auch bei Messungen mit empfindlichsten Geraeten konnte ausserhalb von Bildroehren keine Roentgenstrahlung nachgewiesen werden. Eindeutig feststellbar war hingegen die Strahlung der natuerlichen Radioaktivitaet im Glas der Bildroehre, die auch bei abgeschaltetem Geraet vorhanden ist. Die damit verbundene geringe natuerliche Strahlendosis wird bei jedem Aufenthalt in der Naehe von Glas aehnlich wie vor dem Bildschirm aufgenommen. (SVA-Bulletin (1986) 8)

Spurenanalyse:
LEBERGEWEBE UNTER DER IONENSTRAHLMIKROSONDE

Eine im Kernforschungszentrum Karlsruhe (KfK) entwickelte Ionenstrahlmikrosonde zur hochempfindlichen Konzentrationsbestimmung chem. Elemente in raeumlichen Bereichen mit Abmessungen von wenigen µm wurde jetzt zur Untersuchung des Gehalts von Spurenelementen in krankhaft veraenderten Geweben eingesetzt. Beispielsweise zeigt zirrhotisch - etwa durch uebermaessigen Alkoholgenuss - erkranktes Lebergewebe charakteristische Veraenderungen des Gehalts an Eisen, Kupfer und Brom. Diese Untersuchungen, die vom Institut fuer Kernphysik des KfK zusammen mit der Medizinischen Universitaetsklinik Freiburg und den Stadtischen Krankenanstalten Esslingen durchgefuehrt wurden, eroeffnen neue Wege zum Verstaendnis krankheitsbedingter Stoffwechselstoerungen. Mit der im KfK fuer Materialuntersuchungen entwickelten Mikrosonde wird ein Strahl aus Wasserstoff-, Deuterium- oder Heliumionen mit Energien von bis zu 3,75 MeV, Stromdichte von bis zu 1/100 Ampere pro cm^2 und einem Querschnitt bis herunter zu 3 x 3μm^2 durch ein elektrisches Strahlenfuehrungssystem maeanderfoermig ueber die Oberflaeche der im Vakuum befindlichen Probe gefuehrt. Das Strahlfuehrungssystem gestattet Ablenkungen des Ionenstrahls bis zu 700µm und besonders bei thermisch wenig belastbaren Proben wichtig - lokale Bestrahlungsdauern bis herab zu 100µs. Zum Nachweis des Elementgehaltes der untersuchten Probe existieren im wesntlichen zwei Moeglichkeiten:
Nachweis der protoneninduzierten Roentgenstrahlung mit ihrem fuer die angeregten Elemente charakteristischen Energiespektrum. Dieses Verfahren ist besonders fuer biologische Proben geeignet, da die leichten biologischen Matrixelemente wie Kohlenstoff, Sauerstoff und Wasserstoff nicht, sondern nur die im mittleren Massenbereich liegenden Spurenelemente abgebildet werden.
Nachweis von charakteristischen Produkten wie Protonen und Alphateilchen aus Kernreaktionen zwischen den Atomen der Probe und den hochenergetischen Ionen des Strahls, wobei dieses Verfahren wegen der fuer die Kernreaktionen notwendigen Schwellenenergie nur leichte Elemente, etwa bis zu einer Kernladungszeit von 20, nachweist.
Durch die grosse Vielfalt der bei diesen Verfahren auftretenden Sekundaerstrahlung ist die Mikrosonde auf der Basis von Ionenstrahlen an bestimmte analytische Problemstellungen besser anpassbar als die bekannte Elektronenstrahlmikrosonde. Mit dem Karlsruher System koennen Elementkonzentrationen bis zu 5µg/g Probenmaterial mit einer oertlichen Aufloesung gemessen werden, die die Zuordnung der Konzentrationen zu einzelnen Gewebezellen erlaubt. In Zusammenarbeit mit den medizinischen Instituten wurde jetzt eine Reihe von 60 Lebergewebeproben, darunter aus 10 zirrhotisch veraenderten Lebern, untersucht. Die Proben wurden durch Bauchspiegelung oder bei chirurgischen Eingriffen fuer diagnostische Routineuntersuchungen den Patienten entnommen. Nur ein winziger Bruchteil der Gewebeproben wird fuer die Untersuchung mit der Mikrosonde benoetigt. Die mit der Ionenstrahlmikrosonde nachgewiesenen Spurenelemente sind Bestandteil von wichtigen, den Stoffwechsel regulierenden Eiweissstoffen (Enzymen). Veraenderungen der Enzymkonzentration geben Hinweise auf Stoffwechselstoerungen. Beispielsweise zeigte sich bei einem Patienten mit alkoholtoxischer Leberzirrhose und Eisenmangelanaemie eine deutlische Verminderung des Eisengehaltes. Ein Patient mit primaer biliaerer Zirrhose zeigte dagegen eine homogene Erhoehung der Kupferkonzentration auf das Zehnfache des Normalwertes. Auch wurde erstmals nachgewiesen, dass in solchen Gewebeteilen stark erhoehte Bromkonzentrationen auftraten - neue Hinweise fuer das Verstaendnis der mit der Erkrankung verbundenen Stoffwechselstoerungen. (SVA-Bulletin (1986)8)

Veröffentlichungen / Publications

UNITED KINGDOM ATOMIC ENERGY AUTHORITY

AERE-M 3497
Applications of inductively coupled plasma-mass spectrometry in the nuclear industry -R.M. Brown; S.E. Long; C.J. Pickford, 1985, 19 pp

AERE-M 3519
Publications on sol-gel processing, 1965 to September 1985 - D.L. Segal, 1985, 52 pp.

AERE-R 11141
Sulphate determination in low salinity groundwaters: comparison of results from six analytical techniques - A.R. Lister et al, 1986, 47 pp.

AERE-R 11331
Application of K X-ray fluorescence to the measurement of Pu/U ratios in a variety of solid materials associated with the production of fast reactor mixed oxide fuels - G. Andrew et al, 1985, 24 pp.

AERE-R 11694
Sources of series resistance in the Harwell solid state alpha detector - K.J. Rawlings,

1985, 19 pp.

AERE-R 11915
Radioactive fallout in air and rain: results to the end 1984 - R.S. Cambray et al, 1985, 26 pp.

CLM-R 251
A review of coarse mixing models - D.F. Fletscher, 1985, 35 pp.

CLM-R 252
Assessment and development of the Bankoff and Han coarse mixing model - D.F. Fletscher, 1985, 80 pp.

CLM-R 253 FETRAN
A finite element transport modelling package - Lee - Hsiao. J.W. Eastwood, 1985, 29 pp.

Veranstaltungskalender – Tagungen – Lehrgänge / Calendar – Meetings – Courses

4th Meeting on Nuclear Analytical Methods in Dresden, GDR, May 4 - 8, 1987

The Conference will focus on recent developments in nuclear analytical techniques and methods mainly in the field of -nuclear energetics - industry - geo- and cosmochemistry - microelectronics - medicine and biology - protection of the environment.
There will be plenary lectures, short contributions ans poster discussions presenting original research accomplishments in instrumental and radiochemical activation analysis, nuclear spectroscopy, analysis by nuclear reactions, particle induced x-ray emission, ion-beam techniques, tracer techniques, radiochemical separation techniques, analysis of radioactive materials and nuclear fuels, in-line and automated analysis and the related instrumentation.
Organizing Committee: Dr. sc. A. Hermann, Zentralinstitut fuer Kernforschung Rossendorf, DDR - 8051 Dresden, Postfach 19

13. Weltenergiekonferenz in Cannes

Unter dem Titel "Energie: Bedarf und Erwartung" findet vom 5.-11.10.1986 in Cannes der 13. Kongress der Weltenergiekonferenz statt. Parallel zur Weltenergiekonferenz wird erstmals auch eine grosse technische Fachausstellung zum Thema "Energie: Technik von heute und morgen" durchgefuehrt. Von den Sitzungen und Rundtischgespraechen dieses Kongresses duerfen wie zuvor von den Konferenzen in Istanbul, Muenchen und Neu-Delhi sehr interessante Resultate erwartet werden. Mit der grossen Zahl von 4000-5000 Teilnehmern aus aller Welt bietet die 13. Weltenergiekonferenz ueberdies vielfaeltige Informations- und Kontaktmoeglichkeiten. (SVA-Bulletin (1986) 9)

Ausgewählte Beiträge / Recently published papers

J. RADIOANAL. NUCL. CHEM. ART. 92(1985)2

Trace elements in tissues of a worker affected by rare earths pneumoconiosis - R. Pietra et al

A study of the developmental changes in trace elements and minerals in rat and mouse milk by thermal neutron activation analysis - T. Sato

Comparative study of quality control procedures for Tc-99m radiopharmaceuticals - B. Van Gansbeke et al

J. RADIOANAL. NUCL. CHEM. ART. 97(1986)1

Comparison of the precision and accuracy of the method of sub- and superequivalence and concentration dependent distribution - K. Prikrylovd et al.

Determination of the I-129 content of the primary coolant of nuclear power reactors - I. Kuleff et al

Substoichiometric determination of uranium in phosphate rock - N. Suzuki et al

Detection of weak activities by measurement of gamma-ray spectra - V. Hnatowicz

The status of trace and minor elements in pulses : a PIXE measurement - M. Ali et al

Activation analysis of I-129 - V.J. Filistovich et al

The effects of ionizing radiations on synthetic organic ion exchanges - K.K.S. Pillay

Analysis by nuclear reactions and activation - E. Bujdoso

J. RADIOANAL. NUCL. CHEM. ART. 97 (1986) 2

Alizarin complexone dye solutionsfor radiation dosimetry - N.B. El-Assy et al

Combination of neutron activation, atomic absorption spectroscopy and conventional - M.A. Gouveia et al

Neutron activation analysis of thorium after separation of Th-233 by isotopic ion exchange - C.S. Munita, L.T. Atalla

Radioimmunoassay of somatotropin using second antibody polyethylene glycol combination separation system - G. Samuel et al

Particle size effects in beta-backscattering analysis: grain size effect in loose samples - R. Burek

Automatic activation analysis of fertilizers and plant samples by fast neutrons - S.I. Kafala et al

Multielement analysis of a Milanese air-dust sample by 14 MeV neutron activation - B.M. Bahal, R. Pepelnik

Determination of uranium in sea water by X-ray fluorescence spectroscopy - M. Nagj et al

Analysis by absorption and scattering of radiation - E. Bujdodó

J. RADIOANAL. NUCL. CHEM. ART. 98 (1986) 1

Extraction equilibrium of uranium (VI) with dicyclohexano-18-crown-6 and its application for separating uranium and thorium - Wen-Ji Wang et al

An automatic gamma-ray spectrometer equipped with a micro-robot for sample changing - M. Yagi et al

Interaction of radium with freshwater sediments and their mineral components, III - P. Benes et al

Suppression of charge-induced yield increases in PIGE analysis of biological specimens - A.C. Xenoulis et al

Production of Tc-96 and Tc-95m via the Nb-93 (α,xn) reactions - N. Ramamoorthy et al

An improved method for the determination of Sr-90 in large samples of seawater - J. Borcherding, H. Nies

Some problems associated with the use of boron carbide neutron filters for reactor epithermal neutron activation analysis (ENAA) - F. Chisela et al

Quality control and statistical analysis of biodistribution of commercial Tc-99m-IDA derivatives - V. Jovanovic et al

Isotopic sources in neutron activation analysis - A.N. Garg, R.J. Batra

Environmental radiochemistry and radioactivity - E. Bujdoso

J. RADIOANAL. NUCL. CHEM. ART. 98 (1986) 2

On some response characteristics of high-resolution gamma-ray spectrometers at high count rates - S. Sterlinski et al

A gas-jet system for the study of short-lived, light fission products - K. Rengan, H.C.Griffin

Trace multielement analysis in boron materials by epithermal neutron activation analysis - J.A. Davies et al

The sub- and super-equivalence method of isotope dilution, XI - J. Klas

J. RADIOANAL. NUCL. CHEM. ART. 99 (1986) 1

Separation of lanthanum, cerium and samarium from uranium samples - M. Cassir et al

The limits of decision, detection, and determination in anti-Compton gamma-ray spectrometry - H.A. Das

Ion decontamination with a mixture of cationic and anionic surfactants - N. Filipovic-Vincekovic

Studies on plutonium (IV) hydroxosuccinate - M.S. Nagar et al

Correcting for uranium fission in instrumental neutron activation analysis of high-uranium rocks - M.D. Glascock et al

Neutron activation in geochronology: the Ar-40/Ar-39 technique - M. Oddone, J.M. Villa

Choise of the best region for channel-by-channel summation of estimating the intensity of single photopeakts - L.A. Smakhtin

Evaluation of thick target PIXE analyses with a minicomputer - H. Frey, J. Vogt, G. Otto

Rapid pretreatment of environmental samples of food and related materials by a microwave dehydration method - A. Nakaoka et al

Analysis by nuclear reactions and activation - E. Bujdodo

J. RADIOANAL. NUCL. CHEM. ART. 99 (1986) 2

Solvent extraction of uranium (VI) with di-2-ethylhexyl-dithiophosphoric acid from aqueous nitrate, chloride, sulfate and phosphate media - I. Haiduc et al

Determination of uranium in wet phosphoric acid by extraction with octylphenylphosphoric acid - Z. Ilic, S. Milonjic

Investigations towards an improved Marinelli beaker for gamma detectors - J.D. Hemingway

An external helium flushed milli-beam for combined PIXE and PIGE analysis - A.Hautojaervi et al

Method of separating uranium from iron and thorium - H. Movaseghi

Spectrophotometric determination of uranium (VI) with 1-(2-thiazolylazo)-2naphthol in the presence of Tritan X-100 - G.V. Rathaiah et al

Radionuclide concentrations in several seaweeds and their annual variations - N. Momoshima, Y. Takashima

Fallout of thorium and uranium isotopes from the 1982 eruption of El Chichon volcano - T. Barbod et al

Potentiometric determination of uranium in phosphate-perchlorate medium and its applicability in the natural uranium fuel cycle - A.K. Pandey et al

Analysis by absorption and scattering of radiation - E. Bujdoso

J. RADIOANAL. NUCL. CHEM. LET. 96 (1985) 6

Chemical behavior of microamount of radium transferred from liquid to solid phase - R.S. Wang et al

Nitrogen determination using the N-14(p,α) C-11 reaction at low energies - J.R. Morales, M.I. Dinator

An improced method for purifying α-metallophthalocyanines for use as radionuclide preparatio targets - I.C.S.F. Jardim et al

First use of millicurie levels of Hf-172-Lu-172 in the industrial sector - P.M. Grant et al

J. RADIOANAL. NUCL. CHEM. LET. 103 (1986) 2

A conductivity method for combined carbon and free carbon determination in uranium carbide - G.H. Rizvi et al

Radioimmunoassay of Human Follicle Stimulating Hormone (HFSH) - M. Pimpalkhute et al

J. RADIOANAL. NUCL. CHEM. LET. 103 (1986) 3

The influence of iodine on radiolysis of Purex-solvent - M. Nowak, Z. Nowak

Activation of uranium metal - G.L. Silver

J. RADIOANAL. NUCL. CHEM. LET. 103 (1986) 4

Electrodeposition of uranium and thorium - Y.A. Elewady

Determination of the half-lives of transactinium isotopes by relative activity method - S.K. Aggarwal, H.C. Jain

Epithermal neutron activation analysis. Some comments - Wei-Zhi Tian

Natural radioactivity of building materials - E. Tamez et al

Routine determination of mercury, arsenic and selenium by radiochemical NAA - I. Drabaek et al

ATOMNAJA ENERGIJA 61(1986)1 (all in Russian)

Operating experience and prospects for French Nuclear Power Plants - J. Leclercq

Stage centrifugal extractors - G.I. Kuznetsov

A fast-acting algorithm for calculating neutron field in a reactor operating in a computer dialogue regime - A.M. Afanasyev

Neutron production induced by a bean of high-energy nuclei - I.B. Vorobyov

On the use of a spark counter of fragment tracks for measuring differential fission cross-section - V.E. Rudnikov et al

Some reactions involving the interaction of implanted ions with the surface of solids - M.I. Guseva, G.V. Gordeyeva

Semiconductor telescopic systems - R.A. Muminov et al

Algorithmal accounting of borehole conditions in gamma-ray spectrometry of rocks - D.A. Kozhevnikov

Nuclear-physical characteristics of natural radionuclides - E.M. Krisyuk

Dosimetry of intense electron beams - V.N.Popov

3. Results and Discussions

3.1. *Effect of Medium on the Labelling Yield*

Pure normal alchols. The labelling yield obtained at different time intervals when using as solvent pure normal alcohols ranging in their hydrocarbon chain length, from 1 to 5 carbon atoms (methanol to n-pentanol), was found to be considerably low (Tab. 1). As suggested by many authors [12–14] the main free path for the photochemically induced isotopic exchange between iodo aromatic compounds and radioactive iodine, could be summarised in the following steps:

$$Ar - I \rightarrow [Ar - I]_{ex} \rightarrow Ar. + .I \quad (1)$$

$$I_2^* \rightarrow .I^* + .I^* \quad (2)$$

$$[Ar - I]_{ex} + .I^* \rightarrow Ar - I^* + .I \quad (3a)$$

$$Ar. + .I^* \rightarrow Ar - I^* \quad (3b)$$

where $[\ ]_{ex}$ refers to an excited state.

Tab. 1. Labelling Percent Yield of DIF with Iodine-131 at Different Time Intervals Using Different Solvents

Solvent	D_{50}^*	Yield, % 5 min	15 min	30 min	45 min	60 min	120 min	180 min
Methanol	28·00	7	9	13	19	23	36	38
Ethanol	21·00	8	10	14	20	27	36	40
n-Propanol	16·40	4	6	11	17	20	25	30
n-Butanol	13.85	3	5	7	12	16	20	22
n-Pentanol	11·75	3	5	9	–	17	23	24

D_{50} = dielectric constant at 50 °C

The low yield obtained could be attributed mainly to that in such solvents the main species found are Na^+ and I^-, while as explained in the previous mechanism, the presence of free radioiodine atoms is necessary for the photo-induced isotopic exchange labelling to proceed.

In the pure solvents used the main interaction is an ion-molecule one, which according to *Laidler* and *Eyring* [15] could be enhanced with the decrease in the dielectric constant of the medium. Such postulation could not be verified by Tab. 1, where the exchange yield was found to be relatively higher in case of lower alcohols having higher dielectric constant, than those having lower dielectric constant. However, since the main mechanism in the photo-induced isotopic exchange is a free radical one, the dielectric constant is not the governing factor and could not explain the variation in the exchange yield. The lower exchange yield in higher alcohols may be due to that the longer hydrocarbon chains may act as a better screen which hinder the approach of the solvated reacting species.

Mixture of ethanol and water. Owing to the low solubility of DIF in water and to its higher solubility in ethanol a mixture of both in different ratios were chosen as solvents, and a study has been done to investigate the effect of presence of water on the isotopic exchange rate. From Figs. 1 and 2 it could be observed that the labelling yield increased with increase of water percent (up to 20% water, above which DIF starts to precipitate).

The enhancement of the photo-induced isotopic exchange rate with the presence of water (20%) could be understood, if one takes into consideration the mechanism proposed by *Hayon* [16] which describes the effect of ultraviolet waves on iodide ions and which could be summarized in the following steps:

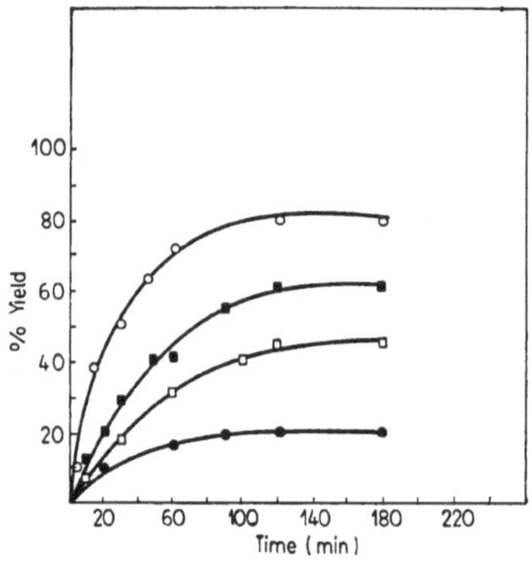

Fig. 1. Variation of the labelling yield as a function of time of exposure using a mixture of ethanol + water in different ratios as solvents (50 mg DIF + 5 ml solvent + 0·1 mCi $Na^{131}I$) Temp. = (30 ± 2) °C
–○– 4 ml ethanol + 1 ml H_2O, –■– 4·5 ml ethanol + 0·5 ml H_2O, –□– 4·75 ml ethanol + 0·25 ml H_2O, –●– 5 ml ethanol

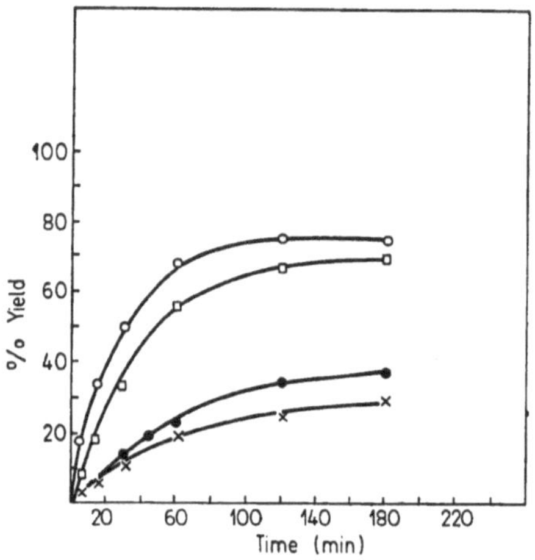

Fig. 2. Effect of presence of water on the variation of the labelling yield as a function of time in methanol and propanol
(50 mg DIF + 5 ml solvent + 0·1 mCi $Na^{131}I$ + 20 λ H_2O_2) Temp. = (30 ± 2) °C
–○– 4 ml methanol + 1 ml H_2O, –●– 5 ml methanol, –□– 4 ml n-propanol + 1 ml water, – × – 5 ml n-propanol

$$I^- H_2O \xrightarrow{h\nu} [I^- H_{20})_{ex} \quad (4)$$

$$[I^- H_2O]_{ex} \rightarrow I^-H_2O + \text{heat} \quad (5)$$

$$[I^-H_2O]_{ex} \rightarrow .I + H_2O^- \quad (6)$$

$$[I^-H_2O]_{ex} \rightarrow .I + H + OH^- \quad (7)$$

$$H + .I \rightarrow H^+ + I^- \quad (8)$$

$$.I + .I \rightarrow I_2 \quad (9)$$

$$H + I_2 \rightarrow .I + I^- + H^+ \quad (10)$$

$$.I + I^- \rightarrow .I_2^- \quad (11)$$

$$2\ .I_2^- \rightarrow I_2 + 2\ I^- \quad (12)$$

On exposure to ultraviolet waves, and in the labelling process the radioiodine molecules or free atoms interact with DIF according to equations (1, 2 and 3). Also it has been suggested that iodoaromatic compounds react very rapidly with hydrated electrons giving iodide and organic radicals [17]:

$$Ar - I + e^-_{aq} \xrightarrow{h\nu} Ar - I^- \rightarrow .Ar + I^- \qquad (13)$$

The organic radical are efficiently scavenged by radio-iodine atoms according to equation (3b).

3.2. Effect of Presence of Hydrogen Peroxide

The photo-induced isotopic exchange rate was found to increase by addition of 20 λ H_2O_2 (30% V/V) to the reacting mixture (4 ml C_2H_5OH + 1 ml H_2O + 50 mg DIF + 20 λ H_2O_2 + 0·10 mCi NaI^{131}) at pH 5·3, Fig. 3. In presence of H_2O_2, formation of I_2 was enhanced according to the following reactions:

$$I + H_2O_2 \rightarrow .I + OH^- + .OH \qquad (14)$$

$$I^- + .OH \rightarrow .I + OH^- \qquad (15)$$

$$2\,.I \rightarrow I_2 \qquad (16)$$

The iodine formed may react with hydrogen peroxide according to the following equation:

$$I_2 + H_2O_2 \rightleftharpoons 2\,HI + O_2 \qquad (17)$$

However, this reaction is infinitely slow in acidic medium.

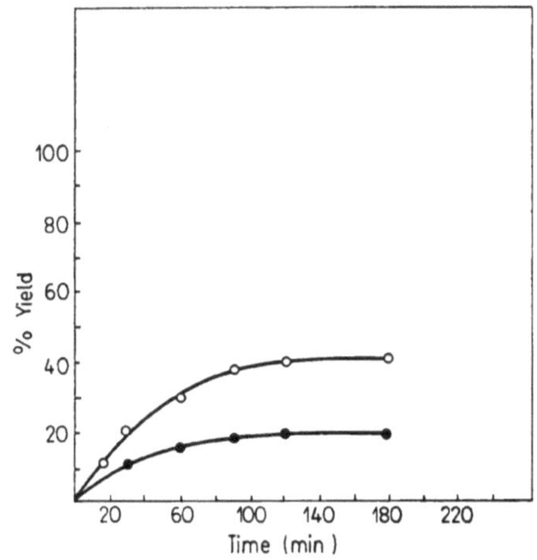

Fig. 3. Variation of the labelling percent yield as a function of time with and without H_2O_2

50 mg DIF + 5 ml ethanol + 0·1 mCi $Na^{131}I$ + λ 20 H_2O_2 (30%) Temp. (30 ± 2) °C

–○– with H_2O_2 (20 λ 30%), –●– without H_2O_2

3.3. Effect of pH of the Medium

By variation of the pH of the medium by using as a solvent 4 ml C_2H_5OH + 1 ml XN NaOH (where X is varied in each experiment), the photo-induced labelling yield was found to decrease with the increase of pH (Fig. 4). This could be explained as being due to that at lower pH values (≈ 5) the oxidation of I^- to I_2 (even in absence of U.V. waves) is rapidly enhanced. At high pH values (alkaline media) the rate of formation of I_2 decreased and in presence of H_2O_2 reaction (19) was to be very fast and leads to the reduction in I_2 concentration. As a matter of fact it has been found that in a slightly alkaline media a yield of 60% was attained after 2 hours irradiation. The mechanism of irradiation of a neutral or alkaline solutions of iodide in presence of H_2O_2 as stated by *Hayon* [16] are:

$$H + H_2O \xrightarrow{h\nu} H_2O + .OH \qquad (18)$$

$$.OH + I^- \rightarrow OH^- + .I \qquad (19)$$

$$I^- + .I \rightarrow I_2^- \qquad (20)$$

$$I_2^- + H_2O_2 \rightarrow H^+ + 2\,I^- + HO_2 \qquad (21)$$

$$HO_2 + HO_2 \rightarrow H_2O_2 + O_2 \qquad (22)$$

In acid solutions, reaction (21) is reversed.

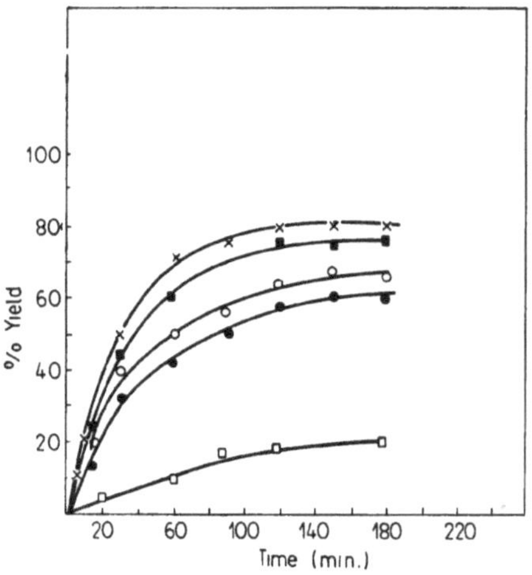

Fig. 4. Variation of the labelling yield % as a function of time at different pH values of the medium

(50 mg DIF + 4 ml ethanol + 1 ml XN NaOH) Temp. = (30 ± 2) °C
– × – pH = 5·2, –■– pH = 6·1, –○– pH = 6·8, –●– pH = 7·6, –□– pH = 8.8

3.4. Effect of Carrier Concentration

The photo-induced isotopic exchange yield was found to increase first with increasing carrier concentration, then it starts to decrease with further increase of the carrier (KI), (Fig. 5). Since, the labelling mechanism is of a free radical type, one would suggest that according to the scavengers experiments done by *Fessenden* and *Schuler* [18], oxygen present in the solution and iodine are competitive scavengers for organic radicals. Therefore, oxygen would compete for the organic radical together with iodine according to the following equations:

$$Ar. + I_2^* \rightarrow Ar - I + .I^* \qquad (23)$$

$$Ar. + O_2 \rightarrow Ar - O_2\,. \qquad (24)$$

There is a small rate-reducing effect of the oxygen being dissolved in the medium. It has been shown by *Fessenden* and *Schuler* [18] that the reducing rate effect of dissolved oxygen should decrease and finally become negligable with increase of iodine concentration. In this way the first increase of isotopic exchange yield with addition of carrier has been explained. The gradual decrease of the photo-

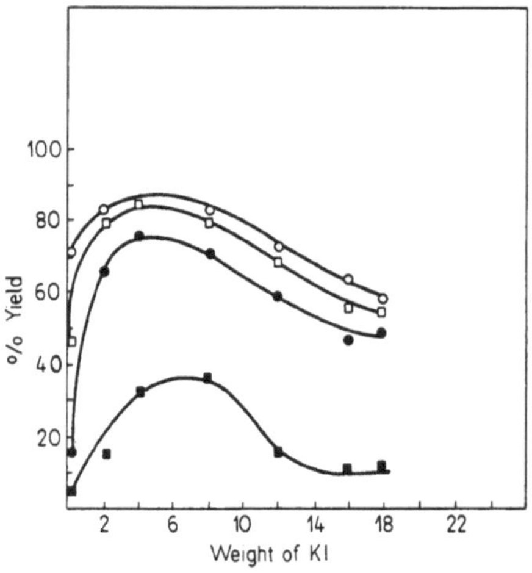

Fig. 5. Effect of amount of carrier on the labelling yield % of different irradiation times

(50 mg DIF ± 4 ml ethanol + 1 ml 0·0298N NaOH + 0·1 mCi $Na^{131}I$), pH = 5·2, Temp. = (30 ± 2) °C
–■– 5 min, –●– 15 min, –□–30 min, –○–60 min

induced isotopic exchange yield with further increase of carrier concentration is mainly due to the isotopic dilution effect.

3.5. *Effect of the Different Lines in the Mercury Lamp Spectra*

The photo-induced isotopic exchange yield obtained when subjecting the reacting mixture to the whole spectra of the Hg lamp was compared with those when different filters are placed between the Hg lamp and the reacting tube (Fig. 6). From the figure it could be observed that the photo-induced exchange rate when using the filter 6543 (absorbs visible and transmits 0% 2300 Å, 45% 2500 Å, 90% 3000 Å, and 0% 4200 Å) is slightly less than when using no filter. When using the filter 16223 B (absorbs visible and transmits 0% 3000 Å, 55% 3600 Å and 0% 4100 Å) a pronounced decrease in the photo-induced exchange rate was observed. A lower exchange rate was obtained when the filter 6544 (absorbs maximum visible and transmits 0% 3100 Å, 38% 3600 Å and 0% 3990 Å) was used.

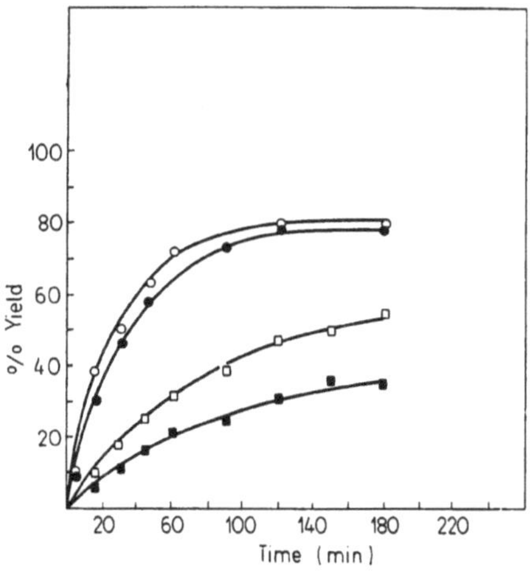

Fig. 6. Variation of the percent labelling yield as a function of time while using different filters
(50 mg DIF + 1 ml H_2O + 4 ml ethanol + 0·1 mCi $Na^{131}I$), Temp. = (30 ± 2) °C
–○– without filters, –●– filter 6543, –□–filter 16223B, –■– ilter 6544

It is clear that the most effective lines in this process are those lying between 2400 Å and 3000 Å taking into consideration that the line 2357 Å in high pressure Hg lamps is reversed. This line is responsible for raising Hg vapour atoms from 6^1 S ground state to the 6^3 P_1 energy level. The shorter wave length lines particularly the 1840 Å line are absorbed by the air gap and by the quartz.

The authors wish to express their thanks to *A. Abdel-Mohty* for his help during this work.

Received November 11, 1985

References

[1] *C. Chiotan, I. Sirbu, I. Zamfir, C. Rentea*, Rev. Chim. (Bucharest **14** (1963) 173
[2] *G. E. Moore*, Science **106** (1947) 130
[3] *G.E. Moore*, Diagnosis and Localization of Brain Tumors. A Clinical and Experimental Study Employing Fluorescent and Radioactiv Tracer Methods, Springfield (Ill. Charles C. Thomas,. 1953
[4] *G. Paroli, A. Zacutti, G. Turchetti, M. F. Becchini*, Minerva Nucleare **7** (1963) 135
[5] *G. V. Taplin, O. M. Meredith, jr., H. Kade, P. Peel, R. Devick* J. Lab. Clin. Med. **45** (1955) 665
[6] *J. Rálková*, Proc. Sec. Internat. Conf. on the Peaceful Uses of Atomic Energy, Geneva: United Nations, 1958, Vol. **20** 142
[7] *G. I. Gleason*, J. Biol. Chem. **213** (1955) 837
[8] *E. Hallaba, E. Abdullah*, Internat. J. Appl. Radiat. Isotopes **17** (1966) 425
[9] *R. Henery*, C.E.A. Report No. 1784 (1960)
[10] *M. Raieh, S. El-Bayoumy, A. Abdel-Mohty*, Isotopenpraxis **17** (1981) 371
[11] *A. Abdel-Mohty*, Labelling of Some Organic Compounds with Radioisotopes, M. Sc. Thesis, Cairo University 1984
[12] *R. K. Sharma, N. Kharash*, Angew. Chem. **80** (1968) 69
[13] *B. Milligan, R. L. Bradow, J. E. Rose, H. E. Hubbert, A. Roe*, J. Amer. Chem. Soc. **84** (1962) 158
[14] *B. Miller, C. Walling*, J. Amer. Chem. Soc. **79** (1957) 4187
[15] *S. Glasstone, K. J. Laidler, H. Eyring*, The Theory of Rate Processes, New York: Mc Graw Hill Book Comp. 1941, p.439
[16] *E. Hayon*, J. Physic. Chem. **65** (1961) 1937
[17] *A. Appleby, J. C. Charlton, R. E. Spillett*, Chem. Comm. **18** (1967) 958
[18] *R. W. Fessenden, R. H. Schuler*, J. Amer. Chem. Soc. **79** (1957) 273

Isotopenpraxis **22** (1986) 8, S. 279

Buchbesprechung

Technik-Wörterbuch Strahlenschutz, Strahlenbiologie, Nuklearmedizin
Englisch – Deutsch – Französisch – Russisch

R. Sube. Berlin: VEB Verlag Technik 1985. 1. Auflage 475 S., 12000 Wortstellen; geb. M 48,–

Das vorliegende Wörterbuch erweitert die bewährte Reihe von fachlich orientierten Wörterbüchern des Verlags Technik um das Fachgebiet Strahlenschutz, Strahlenbiologie und Nuklearmedizin. Was man von einem solchen Wörterbuch erwartet, hängt sicher in hohem Maße von der fremdsprachlichen Qualifikation des jeweiligen Nutzers ab. Außerdem wird die Entscheidung, ob man einen Begriff in einem solchen Wörterbuch erwarten kann oder nicht, umso schwieriger, je spezieller, je eingegrenzter das für die Nutzung vorgesehene Gebiet ist. Von den drei angegebenen Fachgebieten wird am ehesten der Strahlenbiologe trotz der 12000 Wortstellen Lücken empfinden.

Dem Autor kann ohne Einschränkung für das Anliegen des Wörterbuchs und für die Arbeit des Zusammentragens der Vielzahl fachspezifischer Begriffe hohe Anerkennung ausgesprochen werden. Die ausführliche Darstellung im viersprachigen Teil, die übersichtliche Numerierung der Begriffe und die Wortregister in deutsch, französisch und russisch machen das Wörterbuch nutzerfreundlich, übersichtlich und sowohl beim Studium von Fachtexten als auch beim Entwurf von Originalarbeiten in einer Fremdsprache zu einem wertvollen Hilfsmittel.

G. Hübner

Isotopenpraxis 22 (1986) 8, S. 280 bis 284

Verbessertes mikrokatalytisches Verfahren zur Herstellung ^{14}C-markierter Kohlenwasserstoffe

U. Kosakowsky, R. Russow

(VEB „Otto Grotewohl“ Böhlen, KB des PCK Schwedt)[1])

Mit Hilfe des mikrokatalytischen Impulsregimes mit Produktrecycling in Verbindung mit den heute verfügbaren hochaktiven und -selektiven Katalysatoren lassen sich heterogen-katalysierte kohlenwasserstoffwandelnde Reaktionen für eine sehr effektive Herstellung von ^{14}C-Kohlenwasserstoffen nutzen, wie die Erstanwendung dieses Prinzips bei der Gewinnung von [U-^{14}C]Methylcyclopentan zeigte.
Berichtet wird über die Weiterentwicklung dieses Verfahrens durch Erweiterung auf Mehrkomponenteneinsatzprodukte und/oder mehrstufige Reaktionen, d. h. die Kombination verschiedener Reaktions-Trennzyklen in einer Verbundapparatur. Dadurch ist die gleichzeitige Gewinnung mehrerer ^{14}C-Kohlenwasserstoffe nebeneinander in einer Apparatur aus einem ^{14}C-Einsatzprodukt möglich. Die Verfahrensweise wird an der Herstellung von [ring-U-^{14}C]Toluen, [ring-U-^{14}C]Methylcyclohexan und [G-^{14}C]Dimehtylcyclopentan näher erläutert.

As is shown by the first application of this principle to the preparation of ^{14}C-methylcyclopentane heterogeneously catalyzed hydrocarbon converting reactions may be used for the highly efficient production of ^{14}C-hydrocarbons using the microcatalytically pulsed mode of operation with product recycling in connection with the now available highly active and selective catalysts.
The further development of this procedure by its extension to multicomponent feedstocks and/or multistep reactions i.e. to the combination of several reaction and separation cycles in one compound apparatur is reported on. This allows the simultaneous production of several ^{14}C-hydrocarbons in parallel in one apparatus from one ^{14}C-feedstock. The procedure is discussed in more. detail for the preparation of [ring-U-^{14}C]toluene, [ring-U-^{14}C]methylcyclohexane and [G-^{14}C]-dimethylcyclopentane

Keywords

aromatics; carbon 14 catalysis; chemical preparation; cycloalkanes; gas chromatography; hydrocarbons; labelled compounds; labelling

1. Einleitung

Die erstmalige Anwendung des von uns entwickelten mikrokatalytischen Impulsverfahrens für aktiv-präparative Zwecke erfolgte 1979 mit der Herstellung von [U-^{14}C]-Methylcyclopentan aus [U-^{14}C]Cyclohexan oder später [U-^{14}C]Benzen [1, 2].

Dieses Verfahren verwirklicht die Dosierung der Ausgangsverbindung, deren katalytischen Umsatz unter günstigen kinetischen und thermodynamischen Bedingungen (Temperatur, Druck), die gaschromatographische Trennung des Reaktionsgemisches, die Abtrennung des Zielproduktes und den erneuten Einsatz des nicht umgesetzten Einsatzproduktes (Recycling) in einem on-line-gekoppelten Gerätesystem.

Das Prinzip dieser Methode wurde später auch von anderen Autoren auf die Herstellung weiterer ^{14}C-Kohlenwasserstoffe übertragen [3, 4].

Mit dieser einfachen Variante des mikrokatalytischen Impulsverfahrens lassen sich jedoch nur solche Reaktionen für aktiv-präparative Zwecke nutzen, die aus einem Ausgangskohlenwasserstoff in einer einstufigen Reaktion zum Zielprodukt führen (z. B. Hydrierung, Isomerisierung).

[1]) Anschrift: DDR-7202 Böhlen

2. Weiterentwicklung des Verfahrens

Mit der Weiterentwicklung unseres Verfahrens soll eine universelle Anwendung, insbesondere auch die Nutzung von Reaktionen mit zwei und mehr Reaktanten und die gleichzeitige oder alternative Herstellung mehrerer Zielprodukte aus einem Ausgangsprodukt erreicht werden. Bei der apparativen Realisierung erfordert das die Möglichkeit des Einsatzes von Mehrkomponentenprodukten und der mehrstufigen Reaktionsführung. Letzteres, d. h. der Ablauf mehrerer Reaktionen nacheinander (Mehrstufenreaktionen) ist dabei durch eine zweckentsprechende Kombination verschiedener geeigneter Katalysatoren möglich. Abb. 1 zeigt die dafür entwickelte und zur Herstellung der später genannten ^{14}C-Kohlenwasserstoffe benutzte Verbundapparatur. Diese gliedert sich in die Hauptfunktionsgruppen:

1. Dosiervorrichtungen (5, 10), bestehend aus Gasdosierhähnen der Fa. Chromatron und 4-ml-Schleifen (6), gepackt mit 10% Dinonylphthalat auf Porolith, bzw. 30-ml-Schleife ohne Trägermaterial (11)
2. Reaktoren (7) mit Katalysatorfüllung
3. Gaschromatograph (12) mit präparativer Trennsäule und dem Kühlfallensystem (13, 14) zum Abtrennen der Endprodukte.

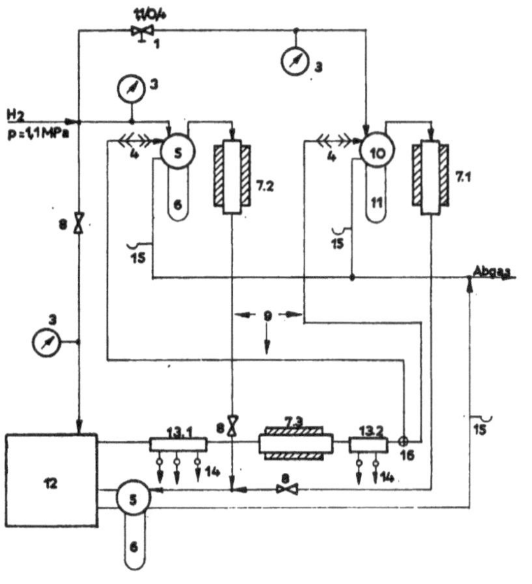

Abb. 1. Schematische Darstellung der Verbundapparatur
1 – Druckminderventil, 2 – Gaseingang, 3 – Manometer, 4 – Ampulle mit ^{14}C-Einsatzprodukt, 5/10 – Gasdosierhähne, 6 – gepackte Dosierschleife 4 ml, 7.1 – 7.3 – elektr. beheizte Reaktoren mit Katalysator, 8 – Nadelventile, 9 – Kapillarleitungen, 11 – Misch- und Dosierbehälter 30 ml, 12 – Gaschromatograph GCAF 18.3 mit präparativer Trennsäule, 13.1 – 13.2 – Verteilerbrücken, 14 – Kühlfallen, 15 – Strömungsmesser, 16 – Dreiwegehahn

Die Reaktoren sind ausgelegt für Temperaturen bis 600 °C. Der Arbeitsdruck für die Gesamtapparatur ist, bedingt durch die benutzten Dosierhähne, bisher noch auf maximal 1,1 MPa begrenzt. Die Kühlung der Dosierschleifen zur Zwischenkondensation und der Kühlfallen erfolgt mittels flüssigem Stickstoff. Das Beheizen der Schleifen zum plötzlichen Verdampfen und damit Dosieren der Einsatz- und Reaktionsprodukte wird durch Niederspannung entsprechender Stromstärke vorgenommen. Nach dem gleichen Prinzip werden die produktführenden Kapillarleitungen beheizt, so daß Kondensationseffekte ausgeschlossen werden können. Nicht umgesetztes Einsatzprodukt und wieder einsetzbare Nebenprodukte können nach der gaschromatographischen Trennung über die beheizte Kapillarleitung in die entsprechende Gasdosierschleife zurückgeführt werden. Dieses Recycling ermöglicht integrale Aktivitätsausbeuten, die mit konventionellen präparativen Synthesen nicht möglich sind. Die Einsatzmengen betragen bei dieser Verfahrensweise in Abhängigkeit von der Reaktion pro Dosierung zwischen 50 und 150 mg. Es lassen sich demzufolge für die Verbindungen hohe spezifische ^{14}C-Aktivitäten erzielen. Alle für die Präparation notwendigen Teiloperationen, nämlich Produktdosierung, katalytische Reaktion und präparative gaschromatographische Trennung der Reaktionsprodukte erfolgen quasi-kontinuierlich nacheinander in diesem on line-gekoppelten Gerätesystem.

Die universelle Anwendbarkeit des von uns entwickelten verbesserten Verfahrens nach dem Prinzip des mikrokatalytischen Impulsregimes sei nachfolgend an Hand des Zielprodukt-Spektrums Alkylaromaten – Naphthene demonstriert.

Ausgehend von den Ergebnissen eigener Untersuchungen zur Toluen-Dealkylierung und -Disproportionierung versuchten wir durch Umkehrung der letzteren Reaktion, also durch eine Synproportionierung von Benzen und Xylen, Toluen zu gewinnen. Die Ergebnisse von gezielten Voruntersuchungen zu dieser Problematik zeigten, daß dieser Weg für die Herstellung von [ring-U-^{14}C]Toluen aus [U-^{14}C]-Benzen nach unserem mikrokatalytischen Impulsverfahren geeignet war und daß sich durch die Konzentration und Art des eingesetzten Methylbenzens (Methylgruppendonator) sowie Reaktionsdruck und -temperatur die Ausbeute an Zielprodukt in bestimmten Grenzen beeinflussen läßt [5].

Auf Grund der gewonnenen Erfahrungen bei der katalytischen Herstellung von [U-^{14}C]Methylcyclopentan in der Gasphase an einem Festbettkatalysator unter erhöhtem H_2-Druck [2] führten wir Untersuchungen zur Übertragung der Ergebnisse auf die Hydrierung/Isomerisierung von Toluen zu Dimethylcyclopentanen und Xylen zu Trimethylcyclopentanen durch. Diese Untersuchungen haben bestätigt, daß unser Verfahren auch dafür geeignet ist. Ausgehend von einem [U-^{14}C]Benzen/Xylen-Einsatzgemisch ergeben sich also die in Abb. 2 schematisch dargestellten Reaktionsmöglichkeiten.

Wie schon erwähnt und aus Abb. 1 ersichtlich, sind Aufbau und Anordnung der Reaktoren so konzipiert, daß diese sowohl separat als auch kombiniert betrieben werden können. Prinzipiell ist es dadurch möglich, mit allen drei Reaktoren einstufige Reaktionen mit einer Einsatzkomponente durchzuführen. Mit der Anwendung einer speziellen Dosiertechnik kann der Reaktor 7.1 auch für den Einsatz und die katalytische Umsetzung von Mehrkomponentenprodukten genutzt werden.

Durch verschiedene Kombinationen der Reaktoren kann die Anzahl und Quantität bzw. Aktivitätsausbeute der Endprodukte gesteuert werden. So sind in Abhängigkeit vom gewünschten Spektrum an Zielprodukten folgende Verfahrensweisen zu wählen:

1. Herstellung von Alkylaromaten
   Nach der Dosierung des Mehrkomponentengemisches sowie der katalytischen Umsetzung am Katalysator des Reaktors 7.1 erfolgt die Abtrennung der gewünschten Alkylaromaten und die Rückführung nicht umgesetzter Einsatzstoffe zu erneuten Reaktion.
2. Herstellung von 6-Ringnaphthenen
   Die Verfahrensweise folgt Pkt. 1, jedoch besteht die

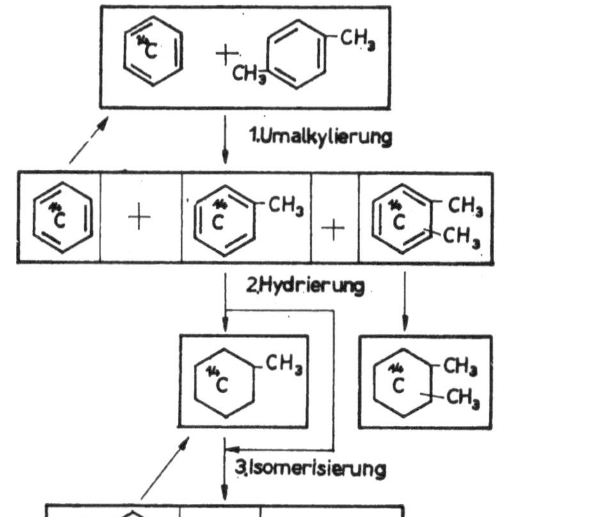

Abb. 2. Reaktionsschema für die Herstellung von [ring-U-^{14}C]Alkylbenzenen, [ring-U-^{14}C]Sechsringnaphthenen und [G-^{14}C]Fünfringnaphthenen

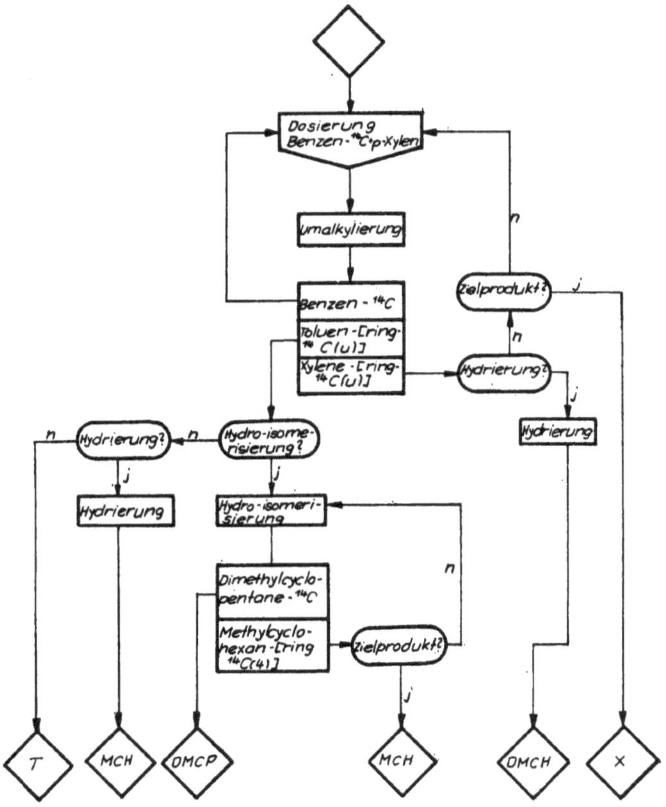

Abb. 3. Programmablauf des beschriebenen Präparationsverfahrens

Möglichkeit, die gebildeten Alkylaromaten am Katalysator des Reaktors 7.3 entweder vollständig zu hydrieren oder abzutrennen, oder nur die Aromaten des ersten Zyklus abzutrennen, und erst die Produkte der folgenden Reaktions-Trennzyklen der Hydrierung zuzuführen. Im letzten Fall können Alkylaromaten und deren Hydrierprodukte nebeneinander gewonnen werden.

3. Herstellung von 5-Ringnaphthenen
Die nach dem unter Pkt. 1 dargestellten Weg gebildeten Alkylaromaten können nach der präparativen gaschromatographischen Trennung entweder abgetrennt oder durch Zuführung des entsprechenden Aromates (C-Zahl= konstant) zum Reaktor 7.2 zu den entsprechenden Naphthenen am Katalysator hydro-isomerisiert werden. Durch Rückführung der zwangsweise auch anfallenden 6-Ringnaphthene (Gleichgewichtseinstellung 5/6-Ringnaphthene) läßt sich die Aktivitätsausbeute an 5-Ringnaphthenen erhöhen. Das Verhältnis von Alkylaromaten, 5- und 6-Ringnaphthenen (Isomerengemische) kann also nach diesem Arbeitsprinzip beeinflußt werden.

Zusammengefaßt sind die möglichen Varianten nochmals in Abb. 3 dargestellt.

3. Praktische Durchführung

Konkret soll die Verfahrensweise am Beispiel der Herstellung von [ring-U-^{14}C]Toluen, [G-^{14}C]Dimethylcyclopentan und [ring-U-^{14}C]Methylcyclohexan aus [U-^{14}C]Benzen erläutert werden.

Für die Präparation dieser Verbindungen werden folgende Katalysatoren und Reaktionsbedingungen gewählt:

1. Stufe: Herstellung von [ring-U-^{14}C]Toluen (Reaktor 7.1)

Katalysator	Na_2O-Mo_2O_3-Mordenit
Reaktionstemperatur:	485 °C ± 5 K
Reaktionsdruck:	0,4 MPa
Katalysatormenge:	1,0 g
H_2-Durchgang:	5,0 l h^{-1}
Einsatzmenge/Dosierung:	150 µl
Einsatzproduktzusammensetzung:	45 Ma.-% [U-^{14}C]Benzen pro 55 Ma.-% p-Xylen
Trennsäule:	3000 mm × 6 mm 15% Dinonylphthalat auf Porolith
Trennsäulentemperatur:	120 °C
H_2-Trägergasstrom:	5 l · h^{-1}
Detektor:	WLZ

2. Stufe: Hydrierung/Isomerisierung von [ring-U-^{14}C]Toluen (Reaktor 7.2)

Katalysator:	Pt/Y-Zeolith (8850)
Reaktionstemperatur:	415 °C ± 5 K
Reaktiondruck:	1,1 MPa
Katalysatormenge:	1,0 g
H_2-Durchgang:	5,0 l · h^{-1}
Einsatzmenge/Dosierung:	110 µl
Trennsäule:	s. 1. Stufe
Trennsäulentemperatur:	70 °C
H_2-Trägergasstrom:	s. 1. Stufe
Detektor:	WLZ

Für diese Präparation werden der Reaktor 7.3 und die Verteilerbrücke 13.2 nicht benötigt.

Vor Beginn der eigentlichen Synthese sind beide Katalysatoren separat zu aktivieren und zu formieren.

1. Na_2O-Mo_2O_3-Mordenit, im folgenden als K 1 bezeichnet (Reaktor 7.1.)

- Trockenen unter N_2-Durchgang
- Aktivieren mit Wasserstoff
- Formieren mit inaktivem Einsatzprodukt

Zur Überprüfung der Katalysatoraktivität werden 150 µl Einsatzprodukt über die Dosiervorrichtung 4 injiziert. Dabei stehen die Dosierhähne 10 und 5 (GC) auf Spülen, und beide Gasschleifen werden gekühlt.

Mittels Feinvakuum am Dosierhahn 10 und durch Erhitzen des Dosierbereiches 4 mit einer Heißluftdusche erfolgt das Übertreiben des Produktes in das Misch- und Dosiergefäß 11. Durch Verschließen und Erhitzen desselben wird das Einsatzprodukt in ein homogenes Gasgemisch überführt und danach in den Reaktor dosiert. Gleichzeitig muß die Kapillarleitung vom Reaktor 7.1 zum Dosierhahn 5 erhitzt werden, die Zuführung vom Reaktor 7.2 ist geschlossen. Nach ca. 60 min befindet sich das gesamte Reaktionsprodukt in der Schleife des Dosierhahnes 5 und kann nun durch Probegabe und Aufheizen der präparativen gaschromatographischen Trennung zugeführt werden. Die Aktivität des Katalysators K 1 ist hinreichend, wenn die Konzentrationen an Toluen ≧ 35 Ma.-% und an Spaltkohlenwasserstoffen ≦ 6 Ma.-% betragen.

2. Pt/Y-Zeolith (8850), im folgenden als K 2 bezeichnet (Reaktor 7.2)

- Dehydratisierung
- Aktivieren mit Wasserstoff
- Formieren mit Toluen

Die Überprüfung der Aktivität des Katalysators K 2 erfolgt analog der des K 1, jedoch mittels Toluen als Einsatzprodukt, das durch Verdampfen unter Systemdruck in den Reaktor 7.2 dosiert wird. Wenn die Konzentrationen der Spaltkohlenwasserstoffe $\leqq$ 3 Ma.-% und das Verhältnis DMCP:MCH $\approx$ 0,7 sind, arbeitet dieser Katalysator hinreichend gut.

Es erfolgt jetzt der Einbau der vorgekühlten, geöffneten und der Einspritzproduktzusammensetzung und -menge (s. 1. Stufe) entsprechend aufgefüllten Ampulle mit [U-^{14}C]Benzen in der Position 4 vor dem Dosierhahn 10 und die Überführung des Produktes in der für K 1 beschriebenen Weise.

Die getrennten Kohlenwasserstoffe des Reaktionsproduktes werden in der Reihenfolge Spaltkohlenwasserstoffe, Benzen, Toluen, Xylene und Trimethylbenzene eluiert und die Spaltkohlenwasserstoffe und Trimethylbenzene in das Abgas geleitet, da eine Aufarbeitung nicht lohnenswert ist. Das nicht umgesetzte Benzen und die Xylene werden über die beheizte Kapillarleitung zum erneuten Einsatz über K 1 in die gekühlte Schleife des Dosierhahnes 10 zurückgeführt und das gebildete Toluen über die beheizte Kapillare in die gekühlte Schleife des Dosierhahnes 5 für K 2 geleitet und gespeichert. In die Dosierschleife für den Reaktor 7.1 erfolgt die Zudosierung der Differenzmengen an Benzen und p-Xylen in der entsprechenden Ausgangszusammensetzung und -menge für den erneuten Reaktions-Trennzyklus. Benzen und Xylene werden danach in der Dosierschleife 11 zwischengespeichert und Toluen mit in die Schleife für Reaktor 7.2 geleitet. Durch Probegabe und Erhitzen dieser Schleife 6 erfolgt die Dosierung des Toluens über K 2 (Dosierhahn des Gaschromatographen auf „Spülen", Schleife gekühlt). Nach ca. 60 min ist die quantitative Überführung des Reaktionsproduktes in diese Schleife beendet, und die Probegabe zur präparativen Trennung wird vorgenommen. Die Spaltkohlenwasserstoffe gelangen durch entsprechende Schaltung des Gasstromes in die Abgasleitung, ^{14}C-Dimethylcyclopentan und ^{14}C-Methylcyclohexan werden im Kühlfallensystem 13/14 abgetrennt. Danach schließt sich wieder ein 2facher Reaktions-Trennzyklus über K 1 und daran ein Zyklus über K 2 an. Dieses „Recycling" wird solange wiederholt, bis ein weiterer Einsatz der Test-^{14}C-Benzen-Aktivität nicht mehr lohnend ist. Bei einer Benzen-Einsatzaktivität von 1570 MBq wären das unter den vorgegebenen Bedingungen 6 Zyklen über K 1 und 3 Zyklen über K 2.

Es ergeben sich dabei folgende Aktivitätsausbeuten bezogen auf eingesetztes U-^{14}C Benzen für

[G-^{14}C]Dimethylcyclopentane:	29%
[ring-U-^{14}C]Methylcyclohexan:	41%
Verluste an ^{14}C:	30%

Nach der hier beschriebenen Arbeitsweise lassen sich prinzipiell auch [G-^{14}C]Trimethylcyclopentane und [ring-U-^{14}C]Dimethylcyclohexane herstellen. Dazu wird zweckmäßigerweise ein [U-^{14}C]Benzen-Trimethylbenzengemisch (40-/60-Ma.-%) eingesetzt, um die Ausbeute an Xylen als Zwischenverbindung zu erhöhen.

Die radiochemische Reinheit der gewonnenen Kohlenwasserstoffe Dimethylcyclopentane und Methylcyclohexan ist > 99% und der ^{14}C-Anteil in der Methylgruppe des Methylcyclohexans < 1%. Ergebnisse über die ^{14}C-Verteilung im Dimethylcyclopentan liegen zur Zeit noch nicht vor.

4. Einschätzung des Verfahrens

Der Vorteil des verbesserten mikrokatalytischen Impulsverfahrens in Verbindung mit der dafür entwickelten Apparatur liegt in der Kombination von drei unabhängig voneinander arbeitenden Mikroreaktoren. Damit haben wir die Möglichkeit, sowohl die im 1. oder 2. Reaktionssystem anfallenden Reaktionsprodukte direkt zu gewinnen (einstufige Reaktion mit Mehrkomponenteneinsatzprodukten oder mit Reinstoff), als auch diese weiteren katalytischen Reaktionen (Mehrstufenreaktionen) zuzuführen. Das „Recycling" erlaubt den optimalen Umsatz des ^{14}C-Einsatzstoffes zum Zielprodukt und eine gewisse Auswahl an zu gewinnenden Endprodukten. Weiterhin bietet dieses Verfahren gegenüber klassischen Arbeitstechniken eine höhere Aktivitätsausbeute und Arbeitsproduktivität. Es ermöglicht ferner eine quasi-kontinuierliche und stabile Arbeitsweise. Durch Anwendung von erhöhtem H_2-Druck lassen sich einige Kohlenwasserstoff-Reaktionen thermodynamisch günstig beeinflussen und die Katalysatordesaktivierung minimieren. Durch die Arbeitstechnik im geschlossenen System wird die potentielle Gefährdung durch Inkorporation von radioaktivem Material verhindert und unkontrollierte Verluste vermieden.

Tab. 1 gibt abschließend eine Übersicht über die von uns nach dem dargestellten Verfahren hergestellten ^{14}C-markierten Kohlenwasserstoffe.

Zum weiteren Ausbau des Verfahrens wäre es sicher entsprechend einer Anregung von *M. Bubner* [6] möglich,

Tab. 1. Nach dem mikrokatalytischen Impulsverfahren hergestellte ^{14}C-markierte Kohlenwasserstoffe

Zielprodukt	Ausgangsprodukt	Q_a
[ring-U-^{14}C]Toluen [ring-U-^{14}C]Xylene	[U-^{14}C]Benzen/p-Xylen	52 21
[ring-U-^{14}C]Toluen [ring-U-^{14}C]Xylene	[U-^{14}C]Benzen/Trimethylbenzen	46 32
[7-^{14}C]Toluen [7,8-^{14}C]Xylene	[7-^{14}C]Benzylalkohol	50 28
[ring-U-^{14}C]Methylcyclohexan	[ring-U-^{14}C]Toluen	90
[ring-U-^{14}C]Dimethylcyclohexane	[ring-U-^{14}C]Xylene	90
[U-^{14}C]Methylcyclopentan	[U-^{14}C]Benzen oder [U-^{14}C]Cyclohexan	64 68
[G-^{14}C]2-Methylpentan [G-^{14}C]3-Methylpentan	[n-1-^{14}C]Hexan	70
[G-^{14}C]Dimethylcyclopentane [ring-U-^{14}C]Methylcyclohexan	[U-^{14}C]Benzen/p-Xylen	29 41
[G-^{14}C]Trimethylcyclopentane [ring-U-^{14}C]Dimethylcyclohexane	[U-^{14}C]Benzen/Trimethylbenzen	in Erprobung
[ring-U-^{14}C]Ethylbenzen	[U-^{14}C]Benzen-/Ethylbenzen	48
[2-^{14}C]2-Methylheptan	[2-^{14}C]2-Methylheptan-2-ol	80

Q_a – Aktivitätsausbeute bezogen auf Einsatzaktivität in %

das hier als Einsatzprodukt verwendete [U-^{14}C]Benzen in einer zusätzlichen, vorgelagerten Reaktionsstufe ebenfalls in der Verbundapparatur zu gewinnen. Es würde sich dafür die katalysierte Trimerisierung von [^{14}C]Ethin, das leicht aus [^{14}C]Bariumcarbid erhältlich ist, anbieten. Entsprechende hochaktive und selektive Katalysatoren sind bekannt (vgl. z.B. [7]). Dadurch ließe sich die Ökonomie des Verfahrens weiter verbessern.

Eingegangen am 2. 12. 1985

Literatur

[1] DDR-WP 149758

[2] *U. Kosakowsky, R. Russow*, Isotopenpraxis **16** (1980) 237

[3] *J. Dermietzel*, Traceruntersuchungen katalytischer Reaktionen von Kohlenwasserstoffen, ZfI-Mitt. Nr. 96

[4] *J. Dermietzel*, J. Lab. Compd. Radiopharm. **18** (1981) 1405

[5] *U. Kosakowsky, R. Russow*, Isotopenpraxis **17** (1981) 382

[6] *M. Bubner*, Kolloquium „Reindarstellung von tritium- und kohlenstoff-14-markierten Verbindungen“, Dresden 1985

[7] *F. Pietig, H. W. Scharpenseel*, Atompraxis **12** (1965) 95

Isotopenpraxis 22 (1986) 8, S. 284 bis 287

Immunological Studies in Diabetics

M. S. Ibrahim[1]), *A. A. Hafiez*[2]), *A. A. Khairat*[3]), *D. Salem*[2])

[1]) Department of Medicine, Al-Azhar University, Cairo, Egypt; [2]) Department of Biochemistry, Cairo University, Cairo, Egypt; [3]) Abdulla Fouad Hospital, Dammam, Saudi Arabia)

Circulating immune complexes (CIC) were detected by a solid-phase Clq-binding radioassay in 13·7% of 73 insulin-dependent diabetics (IDD) as compared to 8·6% of 35 nog-insulin dependent diabetics (NIDD) and 8·3% of 24 control subjects. Islet cell antibodies (ICAb) were detected by the indirect immunofluorescent method in 31·5% of 73 IDD as compared to 8·6% of 35 NIDD and 0% of 24 control subjects. A significant correlation between the incidence of CIC and the incidence of ICAb in IDD was detected and this is in support to the concept that islet cell antibodies are involved in the formation of CIC. The possible role of immuno-pathological mechanism that might involved of IDD is discussed on the base of the significant correlation detected between CIC and ICAb on one hand, and positive diabetic family history, high levels of fasting blood sugar and low levels of fasting insulin level, on the other hand, in Saudi IDD.

Es wurden 108 Diabetiker untersucht, wovon 73 insulinabhängige Diabetiker (IDD) und 35 nicht-insulinabhängige Diabetiker (NIDD) waren. Durch einen Fest-Phasen-Clq-bindenden Radioassay konnten in 13,7% der Fälle von IDD der zirkulierende Immunkomplex (CIC) ermittelt werden, während bei den NIDD nur 8,6% der Fälle CIC bestimmt werden konnte; bei 24 Kontrollpersonen betrug der Anteil 8,3%. Die Inselzell-Antikörper (ICAb) wurden mittels einer indirekten Immunofluoreszenzmethode bestimmt und betrugen 31,5% bei den IDD-Patienten, 8,6% bei den NIDD-Patienten und 0% bei den Kontrollpersonen. Es konnte eine signifikante Korrelation zwischen dem Auftreten von CIC und dem Auftreten von ICAb bei den IDD-Patienten ermittelt werden, was die Annahme unterstützt, daß die Bildung der ICAb mit der Bildung des CIC im Zusammenhang steht. In einer Vergleichsuntersuchung wurden Diabetiker aus Saudi-Arabien und Ägypten untersucht und deutliche Unterschiede festgestellt.

Keywords

antibodies; antibody formation; comparative evaluations; diabetes mellitus; Egyptian Arab Republic; endocrine diseases; immunology; insulin; radioassay; Saudi Arabia

1. Introduction

Since the detection of mononuclear cell infiltrates in the islets of *Langerhans*, of a child who died early in the course of diabetes mellitus, by *von Meyenburg* [1] evidence has been accumulated implicating potential immune mechanisms in the pathogenesis of the disease in insulin-dependent diabetics; where they have lymphocytic infiltration of the pancreatic islets [2] a higher frequency of other organ-specific auto-antibodies, e.g. thyroid and gastric [3], evidence of cell mediated immunity against pancreatic islets [4, 5], circulating organ-specific islet cell antibodies [6, 7, 8], a higher prevalence than in the normal population of certain HLA-antigens [9, 10] and finally a higher prevalence of circulating immune complexes [11, 12, 13, 14].

Address: [2]) Author to whom all correspondence to be addressed; Department of Biochemistry, Faculty of Medicine, Cairo University; Kasr El-Aini, Cairo, Egypt

This last observation prompted us to investigate the incidence of circulating immune complexes in Saudi diabetics and to correlate this incidence to the clinical representation of diabetes, diabetic family history, fasting blood sugar and fasting insulin level; and to other immunological factors as islet cell antibodies. Then, we carried out a comparative study between the incidence of circulating immune complexes in Saudi diabetics and in Egyptian diabetics.

2. Materials and Methods

A total of 108 Saudia diabetics attending the Out-patient Diabetic Clinic of Abdulla Fouad Hospital, Dammam, were studied. The National Diabetes Data Group (NDDG) diagnostic criteria for diabetes 1979 [15] were used to select this group of diabetic patients and to classify them into

Tab. 1. Age distribution of Saudi diabetic patients and controls

Age Group Years	IDDM (n = 73)		NIDDM (n = 35)		Controls (n = 24)	
	No.	%	No.	%	No.	%
10 – 15	2	2·7	–	–	–	–
15 – 20	7	9·6	–	–	4	16·7
20 – 25	11	15·1	–	–	7	29·2
25 – 30	7	9·6	–	–	2	8·3
30 – 35	28	38·3	–	–	2	8·3
35 – 40	18	27·7	–	–	3	12·5
40 – 45	–	–	13	37·1	2	8·3
45 – 50	–	–	12	34·3	2	8·3
50 – 55	–	–	5	14·3	1	4·2
> 55	–	–	5	14·3	1	4·2
Total	73	100·0	35	100·0	24	100·0
Mean age ± S.D.	30·17 6·96		48·23 5·17		30·3 12·1	

Tab. 2. Distribution of Saudi diabetic patients according to the duration of the disease before inclusion in the study

Duration	IDDM		NIDDM	
	No.	%	No.	%
< 1 m	6	8·2	4	11·4
1 – 6 ms	5	6·8	4	11·4
6 – 12 ms	8	11·0	5	14·3
1 – 2 yrs	17	23·3	10	28·6
2 – 5 yrs	27	37·0	6	17·1
5 – 10 yrs	10	13·7	3	8·6
> 10 yrs	–	–	3	8·6
Total	73	100·0	35	100·0
Mean	2·8 years		2·5 years	

Tab. 3. Age distribution of Egyptian diabetic patients and controls

Age Group Years	IDDM (n = 41)		NIDDM (n = 18)		Controls (n = 12)	
	No.	%	No.	%	No.	%
< 5	1	2·4	–	–	1	8·3
5 – 10	12	29·3	–	–	3	25·0
10 – 15	15	36·6	–	–	1	8·3
15 – 20	7	17·0	–	–	3	25·0
20 – 25	2	4·9	–	–	–	–
25 – 30	2	–	–	–	–	–
30 – 35	2	4·9	–	–	1	8·3
35 – 40	2	4·9	1	5·6	–	–
40 – 45	–	–	4	22·2	1	8·4
45 – 50	–	–	1	5·6	1	8·4
50 – 55	–	–	7	38·8	1	8·3
> 55	–	–	5	27·8	–	–
Total	41	100·0	18	100·0	12	100·0
Mean age ± S.D.	14·68 6·9		49·11 3·8		21·3 16·9	

73 insulin-dependent and 35 non-insulin dependent. Their age distribution is shown in Tab. 1, while Tab. 2 shows their distribution according to the duration of the disease before inclusion in the present study. Routine clinical examination and investigations were carried out with special attention to diabetic family history and fasting blood sugar.

Circulating immune complexes were detected by solid-phase Clq-binding radioassay according to the method of *Hay* et al. (1976) [16], while indirect immunofluorescent method as described by *Irvine* et al. (1977b) [17] was the method used for detection of islet cell antibodies; and finally commercial kits (RA systems Laboratories, USA) were used to detect fasting insulin level.

Tab. 4. Distribution of Egyptian diabetic patients according to the duration of the disease before inclusion in the study

Duration	IDDM		NIDDM	
	No.	%	No.	%
< 1 m	1	2·4	–	–
1 – 6 ms	5	12·2	–	–
6 – 12 ms	6	14·6	2	11·11
1 – 2 yrs	11	26·8	2	11·11
2 – 5 yrs	12	29·3	3	16·7
5 – 10 yrs	6	14·7	6	33·3
> 10 yrs	–	–	5	27·8
Total	41	100·0	18	100·0
Mean	2·9 years		7·8 years	

Also a total of 59 Egyptian diabetic patients (41 insulin-dependent and 18 non-insulin dependent) attending the Out-patient Diabetic Clinics of Al-Husain Iniversity Hospital and Kasr El-Aini Children Hospital, Cairo, were studied for the detection of the incidence of circulating immune complexes using the solid-phase Clq-binding radioassay according to the modification of *Hay* and *Nineham* (1982) [18]. Their age distribution is shown in Tab. 3, while Tab. 4 shows their distribution according to the duration of the disease before inclusion in the present study.

The control groups comprised 24 normal Saudi subjects and 12 normal Egyptian subjects, clinically free, not showing any manifestations of immunological abnormalities and match with the age and sex distribution of their corresponding diabetic patients.

3. Results

Incidence of CIC in Saudi diabetics:

CIC are found with a significantly higher frequency in IDD (13·7%) than in controls (8·3%) and in NIDD (8·6%). As shown in Tab. 5, the incidence of CIC has a bimodal distribution in IDD where the highest incidence of CIC can be seen in diabetics with very short or very long duration of illness while the lowest incidence of CIC can be seen in diabetics with intermediate durations of illness.

Incidence of ICAb in Saudi diabetics:

ICAb are found with a significantly higher frequency in IDD (31·5%) than in controls (0%) and in NIDD (8·6%). Still, we can see that the frequency of ICAb is also significantly higher in NIDD than in controls. Also, Tab. 5 shows that higher frequency of ICAb occurs in diabetics with short duration of the disease and lower frequency occurs with longer durations.

Correlation between CIC and ICAb:

A highly significant association between CIC and ICAb is found in the group of IDD (Tab. 6) but not in NIDD. Eighty per cent of IDD with CIC have ICAb while 34·7% of IDD with ICAb have CIC.

Correlation between CIC and clinical representation of IDD. (Tab. 7) in the form of:

a- Positive diabetic family history (FH), a significantly higher positive diabetic family history, first and second degrees were only considered, is detected among CIC positive IDD than among CIC negative IDD.

Tab. 5. The prevalence of circulating immune complexes, islet cell antibodies and insulin deficiency in both Saudi diabetic groups, distributed according to the duration of diabetes, and controls

Duration	Total No.	CIC (+ ve)		ICAb (+ve)		Insulin Deficiency		
		No.	%	No.	%	No.	%	Mean insulin level u/ml
IDDM						(< 6 μu/ml)		
< 1 m	6	2	33·3	4	66·6	4	66·6	4·9
1 – 6 ms	5	1	20·0	2	40·0	3	60·0	6·3
6 – 12 ms	8	1	15·5	4	50·0	4	50·0	7·0
1 – 2 yrs	17	1	5·8	7	41·0	9	52·9	8·5
2 – 5 yrs	27	3	11·1	5	18·5	12	44·4	6·8
> 5 yrs	10	2	20·0	1	10·0	5	50·0	6·6
Total	73	10	13·7 (P < 0·001)	23	31·5 (P < 0·001)	37	50·7	7·1
NIDDM						(12 μu/ml)		
< 1 m	4	–	–	1	25·00	1	25·00	13·25
1 – 12 ms	9	–	–	1	11·11	2	22·2	14·33
1 – 2 yrs	10	–	–	1	10·0	4	40·0	15·90
2 – 5 yrs	6	1	16·7	0	0·0	6	100·0	10·50
3 – 10 yrs	3	1	33·3	0	0·0	3	100·0	10·00
> 10 yrs	3	1	33·3	0	0·0	3	100·0	8·70
Total	35	3	8·6 (Not significant)	3	8·6 (P < 0·001)	19	54·2	12·7
Control	24	2	8·3	0	0·0	1	4·16	

Tab. 6. Correlation between CIC and ICAb in Saudi Insulin-dependent diabetics

		ICAb		
		(+ ve)	(– ve)	
CIC	(+ ve)	8	2	10
	(– ve)	15	48	63
		23	50	73

Tab. 7. Correlation between CIC and Clinical representation of insulin-dependent diabetes in the form of family history (FH), fasting blood sugar (FBS) and fasting insulin level

CIC	No.	+ve F.H.		F.B.S. (mean level)	Fasting Insulin (mean level)
		No.	%		
(+ve)	10	5*	50·0	206·5** mg/dl ± 49·0	3·45*** μ U/ml ± 1·46
(–ve)	63	9	14·3	159·8 mg/dl ± 19·3	7·70 μ U/ml ± 4·90

* Significantly higher (+ve) diabetic F.H. among (+ve) CIC diabetics ($x^2 = 4{\cdot}55$, $P = 0{\cdot}05$).
** Significantly higher FBS among (+ve) CIC diabetics ($t = 5{\cdot}57$, $P < 0{\cdot}001$).
*** Significantly lower mean fasting insulin level among (+ve) CIC diabetics ($t = 2{\cdot}71$, $P = 0{\cdot}01$).

Tab. 8. The comparative data for both ethic diabetic groups (Saudis and Egyptians)

		IDDM		NIDDM	
		Saudis	Egyptians	Saudis	Egyptians
Mean age (years)		30·17 ±6·96	14·68 ±6·9	48·23 ±5·17	49·11 ±3·8
Mean duration of illness (years)		2·8	2·9	3·5	7·8
(+ve) CIC	No.	10	14	3	2
	%	13·7	34·2	8·6	11·1

b- Fasting blood sugar (FBS)
a significantly higher mean FBS is detected among CIC positive IDD than among CIC negative IDD.

c- Fasting insulin level
a significantly lower mean fasting insulin level is detected among CIC positive IDD than among CIC negative IDD.

Comparative study between the incidence of CIC in both Saudi and Egyptian diabetics.

As shown in Tab. 8, the mean age of Egyptian IDD is significantly different from the mean age of the Saudi IDD ($P < 0{\cdot}001$), but the mean duration of diabetes in both groups matches with each other. Also, a statistically significant higher incidence of CIC is detected in Egyptian IDD relative to Saudi IDD ($P < 0{\cdot}001$).

4. Discussion

The high prevalence of CIC in IDD and the close correlation between CIC and ICAb which were previously reported in Western population [11, 13, 14, 19] has been confirmed by the present study on Saudi population. Nevertheless, the prevalence of CIC in Saudi IDD in our study (13·7%) is less than what has been reported by *Irvine* et al. [19] (39%) and by *Deplespesse* et al. (1980) [20] (34%); this could be explained as follows:

a- We used only one method (solid-phase Clq-binding radioassay) while more than one method differing in principle were used in their studies.

b- This study included Saudi IDD with different durations of illness, from 3 – 4 days upto 7 years, while other studies were mainly on IDD with a duration of illness less than one month; still our incidence of CIC in Saudi IDD during the first month of the disease is 33·3% which coincides with the previously mentioned reports.

c- The prevalence of various complications of diabetes in long standing IDD was higher in previous comparative studies, e.g. 50% of 32 IDD had various diabetic complications in the study of *Irvine* et al. [19] while only 23% of 73 IDD in our study had diabetic complications.

Although the used method of detection of ICI in this study gives no indication of the nature of the antigens involved, the detected correlation between CIC and ICAb together with the comparable decline of the prevalence of both CIC and ICAb with time following diagnosis of the disease stronger suggests that islet cell antigen-antibodies may well contribute to the CIC that occurs in Saudi IDD specially those who were studied early in the course of the disease.

The possible role of immuno-pathological mechanisms that might be involved in the development of insulin-dependent diabetes as has been suggested by *Irvine* [21] has been supported in the present study through the detected correlation between CIC and ICAb in, on one hand, and the detected correlation between high prevalence of CIC and (a) high mean fasting blood sugar, (b) low mean fasting insulin level, on the other hand, in Saudi IDD. Also, the detected lower mean fasting insulin level in Saudi IDD with positive ICAb than in those with negative ICAb is a further support to *Irvin*'s hypothesis.

CIC consisting of islet cell-antigen-antibody may arise as a consequence of islet cell damage caused by some chemical or virus but it is also accepted that such complexes may have a primary role in pathogenesis of islet cell failure.

Such complexes could conceivably "arm" K-cells [22] to render them specifically cytotoxic to islet cells. Alternatively the presence of such complexes within the islets may activate the complement system with the generation of its biologically active components [19]. Again, impaired clearance of CIC from the circulation of the reticulo-endothelial system may also contribute to their presence in the circulation as it has been shown that the phagocytic function is subnormal during metabolic decompensation in diabetes [23].

In the comparative part of this study between Saudi and Egyptian diabetics we are claiming that more immunological influence is detected in Egyptian IDD, than in Saudi IDD. This claim is built on two observations:

a- The prevalence of CIC in Egyptian IDD (34·2%) is significantly higher than in Saudi IDD (13·7, $P < 0{\cdot}0001$), while the method used for detection of CIC in both ethic groups is the same and the mean duration of diabetes in both ethic groups is the same and the mean duration of diabetes in both ethic groups is not significantly different from each other.

b- The mean age of Egyptian IDD (14·68 years) is significantly different from the mean age of Saudi IDD (30·17 years) $P < 0{\cdot}001$. The former group coincide with the early peak for the onset of insulin dependent diabetes while the latter group coincide with the late peak. *Christau* et al. [24] postulated the occurrence of two peaks for the onset of insulin dependent diabetes, while *Cudworth* et al. [25] postulated that diabetics who were diagnosed at the early peak are more susceptible to have two "high risk" HLA-antigens, and according to the attractive hypothesis of *Bloom* et al. [26] and *Ludvigsson* et al. [27] that HLA-linked genes control the immune responsiveness to the factors that have a potential cytoplasmic effect on the B cells in the pancrease, so our claim can be conceivable.

No similar findings was detected among non-insulin dependent diabetics in both ethic groups.

Received November 6, 1985

References

[1] *H. von Meyenburg*, Med. Wochenschr. **21** (1940) 554
[2] *W. Gepts*, Diabetes **14** (1956) 247
[3] *W. J. Irvine, B. F. Clarke, L. Scarth, D. R. Cullen, L. J. P. Duncan*, Lancet **2** (1970) 163
[4] *J. Nerup, O. O. Anderson, G. Bendexen, J. Engberg, J. E. Poulsen*, Diabetes **20** (1971) 424
[5] *A. C. MacCuish, J. Jordan, C. J. Campbell, L. J. P. Duncan, W. J. Irvine*, Diabetes **23** (1974) 439
[6] *G. F. Bottazzo, A. Florin-Christensen, D. Doniach*, Lancet (1974) 1279
[7] *A. C. Mac Cuish, F. W. Barnes, W. J. Irvine, L. J. P. Duncan*, Lancet (1974) 1529
[8] *K. James, S. R. Bruce*, Diabetes **33** (1984) 8
[9] *M. Thomsen, P. Platz, O. O. Anderson, M. Christy, J. Lyngsoe, J. Nerup, K. Rasmussen, L. Ryder, L. Nielsen, A. Suejgaard*, Transplant Review **22** (1975) 120
[10] *J. Nerup, P. Platz, L. P. Ryder, M. Thomsen, A. Suejgaard*, Diabetes **27** Suppl. 1 (1978) 247
[11] *W. J. Irvine, S. F. Al-Kateeb, U. DiMario, C. M. Feek, R. S. Gray, B. Edmond, L. J. P. Duncan*, Clin. Exp. Immunol. **30** (1977) 16
[12] *U. DiMario, M. Iavicoli, D. Andreani*, Diabetologia. **19** (1980) 89
[13] *G. Virella, H. Wohltmann, J. Segal* et al., Diabetologia **21** (1981) 184
[14] *G. Triola, E. Giardina, A. Rinaldi, G. D. Bompiani*, Clin. Immunol. Immunopathol. **30** (1984) 169
[15] National Diabetes Data Group, Diabetes **28** (1979) 1039
[16] *F. C. Hay, L. J. Nineham, I. M. Roitt*, Clin. Exp. Immunol. **24** (1976) 396
[17] *W. J. Irvine, C. J. McCallum, R. S. Gray, G. J. Campbell, L. J. P. Duncan, J. W. Farguhar, H. Vaughn, P D. Morris*, Diabetes **26** (1977) 138
[18] *F. C. Hay, L. J. Nineham*, Personal communication (1982)
[19] *W. J. Irvine, U. DiMario, K. Guy, C. N. Feek, R. S. Gray, L. J. P. Duncan, W. J. Irvine* (Ed.), Teviot Scientific Publications, Edinburgh (1980) p. 219
[20] *G. Delespesse, P. H. Gausset, M. Sarfati, M. Dubi-Rucquay, M. J. Debischop, L. Wanhaelst*, Clin. Exp. Immunol. **40** (1980) 196
[21] *W. J. Irvine, W. J. Irvine* (Ed.), Teviot Scientific Publications, Edinburgh (1980) P. 1, 349
[22] *E. A. Calder, D. McLennan, W. J. Irvine*, Clin. Exp. Immunol. **15** (1973) 467
[23] *D. D. Bagdade*, Acta Endocrinol. **83**, Suppl. 204 (1976) 27
[24] *B. Christau, H. Kromann, O. O. Anderson, M. Christy, U. Buschard, K. Arnung, I. H. Kristensen, B. Pietersen, J. Steinrud, J. Nerup*, Diabetologia **13** (1977) 281
[25] *A. Cudworth, G. F. Bottazo, D. Doniach, W. J. Irvine* (Ed.), Teviot Scientific Publications, Edinburgh (1980) pp. 67 – 96
[26] *A. Bloom, T. M. Hayes, D. R. Gamble*, Britt. Med. J. **58** (1975) 583
[27] *J. Ludvigsson, L. Hedding, J. Safwenberg*, In HLA and Disease, INSERM, Paris (1976) 133

Isotopenpraxis 22 (1986) 8, S. 288 – 291

Anwendung von Radiotracern bei der Entwicklung hydrometallurgischer Verfahren zur Aufarbeitung von Sekundärrohstoffen der NE-Metallurgie mittels Flüssig-Flüssig-Extraktion[1])

T. S. Urbański

(Instytut Badań Jadrowych (Institut für Kernchemie und Kerntechnik)[2])

Mit Hilfe radioaktiver Nuklide wurde die selektive Abtrennung von Cu, Zn und Mn aus verschiedenen flüssigen Sekundärrohstoffen durch Flüssig-Flüssig-Extraktion untersucht. Dabei gestattete es die Versuchsapparatur, die Einstellung des Extraktionsgleichgewichtes in einer Mixer-Settler-Zelle kontinuierlich zu verfolgen. An Hand eines neu eingeführten Selektivitätskriteriums wurde die Eignung verschiedener industrieller Extraktionsmittel (u. a. LIX 64 N, Acorga P-17, Kelex 100, D 2 EHPA) zur Extraktion dieser Buntmetalle aus erschöpften Zink-Elektrolyten bzw. Laugelösung von Cu/Zn-Schrott bestimmt. Gute Ergebnisse konnten durch optimale Wahl der Extraktions- und Rückextraktionsbedingungen erreicht werden.

The selective separation of Cu, Zn and Mn from various liquid secondary raw materials has been studied by the use of radionuclides. The experimental equipment allowed the establishment of equilibrium in a mixer-settler to be continuously followed. The suitability of various industrial extractants such as LIX 64 N, Acorga P-17, Kelex 100, D 2 EHPA for the extraction of those non-ferrous metals from spent zinc electrolytes and from solutions obtained by leaching Cu/Zn scrap has been determined on the basis of a new selectivity criterion. Good results have been achieved by selecting the optimum extraction and reextraction conditions.

Keywords

brass; copper base alloys; copper 64; manganese 54; mixer-settlers; optimization; radioisotopes; reprocessing; separation processes; solvent extraction; tracer techniques; zinc 65

Verschiedene Nichteisenmetalle können durch hydrometallurgische Verfahren, u. a. mittels Flüssig-Flüssig-Extraktion, aus unterschiedlichen flüssigen Sekundärrohstoffen, wie z.B. Lösungen von Metall- und Legierungsabfällen, Laugungslösungen der Metall- und Legierungsaufbereitung, erschöpfte Elektrolyte (z.B. der Zink-, Kupfer- oder Nickel-Elektrolyse), industriellen Abwässern usw., extrahiert werden. Die Wiedergewinnung der Metalle mittels Extraktion kann sich ökonomisch lohnen und gleichzeitig die Belastung der Umwelt vermindert werden.

Bei der Verfahrensentwicklung sowie der Optimierung möglichst selektiver Extraktions- und Reextraktionsprozesse können mit großem Erfolg Radiotracer eingesetzt werden.

Experimentell wurde unter Verwendung der Radiotracer ^{54}Mn, ^{64}Cu und ^{65}Zn die selektive Abtrennung von Kupfer, Zink und Mangan aus synthetischen Lösungen untersucht. Deren Zusammensetzung entsprach der von erschötften Zinkelektrolyten bzw. der Laugelösungen, die bei der Laugung von Cu/Zn-Legierungen mit Ammoniak oder Mineralsäuren erhalten werden. Die Versuchsapparatur (Abb. 1) bestand aus einem thermostatierbaren Mixer-Settler-Extraktor (Arbeitsvolumen für beide Phasen 1 dm³), ausgerüstet mit einem Turbinenrührer sowie Elektroden zur pH-Wertmessung. Die Messung der γ-Strahlen der Radiotracer erfolgte kontinuierlich, indem durch getrennte Leitungen Anteile der organischen bzw. wäßrigen Phase

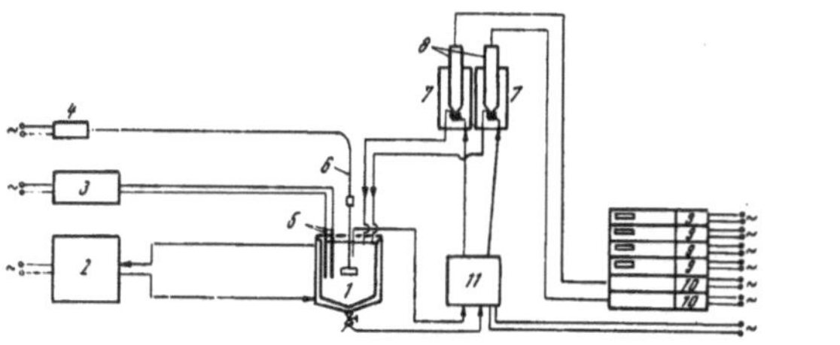

Abb. 1. Schema der Versuchsapparatur:
1 Mischerabsetzerextraktor, 2 Thermostat, 3 pH-Meter, 4 Rührwerkmotor, 5 Elektroden, 6 Rührwerk, 7 Bleischirm, 8 Meßsonden, 9 Einkanalanalysatoren, 10 Stromversorgung mit Verstärker, 11 Peristaltische Mehrfachpumpe

zu den Meßsonden gepumpt und nach der Zählung in den Extraktor zurückgeführt wurden. Beide Meßsonden wurden bei identischer Geometrie mit den jeweils benutzten Radionukliden geeicht, um gleiche Zählgeschwindigkeiten zu erhalten. Ferner wurde geprüft, ob die Summe der in beiden Phasen gemessenen Zählraten stets konstant war. (Bei Verwendung von Spektrometersonden wäre es möglich, zwei Radiotracer gleichzeitig einzusetzen und aus den entsprechenden Meßpunkten sofort Informationen über den Trenneffekt zu erhalten).

Aus den Zählraten der organischen und wäßrigen Phase I_0 bzw. I_w (Mittelwerte von ≥ 4 Messungen) wurde der Extraktionskoeffizient E berechnet:

$$E = \frac{I_0}{I_w}$$

Zur Bestimmung der Selektivität der Trennung im Extraktions- oder Reextraktionsprozeß wurde ein neues

[1]) Vortrag, gehalten auf dem 7. Kolloquium „Technische Anwendung radioaktiver Isotope im Berg- und Hüttenwesen", 13. 3./14. 3. 1985 Hetzdorf b. Freiberg, DDR

[2]) Anschrift: 03-195 Warszawa, ul. Dorodna 16, VR Polen

Komplexkriterium herangezogen, welches die drei folgenden, gleichzeitig bestimmten Größen enthält:

$$a = \lg E_1$$
$$b = \lg (C_1/C_2)_0$$
$$c = \lg (C_2/C_1)_w$$

Index „1" bezeichnet hier das besser extrahierte Metall, „0" – organische Phase, „w" ≧ wäßrige Phase.

Diese Größen bestimmen:

a: die Extraktionsausbeute des besser extrahierten Metalls

b und *c*: die Reinheiten der beiden zu trennenden Metalle im Extrakt (*b*) bzw. im Raffinat (*c*).

Trägt man diese Größen gemeinsam in Abhängigkeit von Werten des interessierenden Prozeßparameters der Extraktion oder Reextraktion auf (z.B. pH-Wert der wäßrigen Phase, Solvenskonzentration u.a.), kann man Bereiche der maximalen Selektivität (MS) erkennen, in denen bei festgelegten *x*-Werten die Größen *a*, *b* und *c* gleichzeitig ihre höchsten Werte erreichen. Man kann aber auch in Abhängigkeit von der Aufgabenstellung Werte für *a*, *b* und *c* vorgeben und dann dem erwähnten Diagramm den Bereich praktisch ausreichender Selektivität (bezeichnet als GS_{abc}) entnehmen, in dem $a \geqq a_{min}$, $b \geqq b_{min}$, $c \geqq c_{min}$ ist. Dies wird unten an praktischen Beispielen gezeigt.

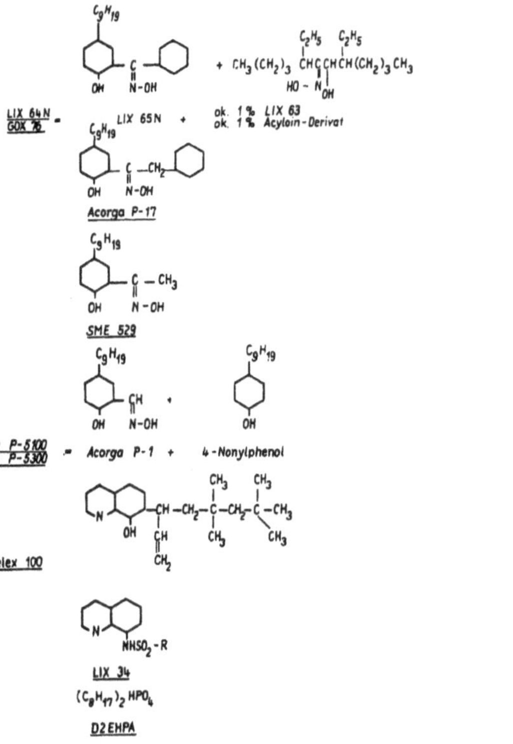

Abb. 2. Zusammenstellung der zur Cu/Zn-Trennung benutzten Extraktionsmittel

Es wurde die Selektivität der extraktiven Kupfer/Zink-Trennung beim Einsatz industrieller Extraktionsmittel untersucht. Es handelte sich um die β-Hydroxyoxime LIX 64 N, GOX 76, Acorga P-17, SME 529, Acorga P-5100 und Acorga P-5300, Alkylderivat von 8-Hydroxychinolin Kelex 100, Sulfonamidderivate und Chinolin LIX 34 und Di-[2-ethyhexyl]-phosphorsäure D 2 EHPA. Die chemischen Strukturen werden in Abb. 2 gezeigt.

Die Selektivität der Kupfer- und Zink-Extraktion aus Ammoniaklösungen mit sechs verschiedenen Extraktionsmitteln ist als Beispiel in Tab. 1 dargestellt. Die Cu- und Zn-Gehalte betrugen 0,01 M, die Solvenskonzentration 0,2 M, entsprechend einem fünffachen stöchiometrischen Überschuß. In allen Fällen sind die *a*- und *c*-Werte hoch, aber die *b*-Werte für Kelex 100 und LIX 34 sind praktisch gleich Null. Das heißt, diese Extraktionsmittel extrahieren Cu und Zn unter den gewählten Bedingungen gemeinsam. Die *b*-Werte sind etwas höher für Acorga P-5100 und P-5300, d. h. diese Extraktionsmittel sind schwach selektiv. Dagegen sind LIX 64 N, SME 529 und auch Acorga P-17 (nicht in der Tabelle gezeigt) selektive Extraktionsmittel. D 2 EHPA im Gegensatz zu den anderen verwendeten Extraktionsmitteln extrahiert Zink besser als Kupfer.

Interessante Ergebnisse zur selektiven Trennung kann man im Reextraktionsprozeß erzielen, auch wenn das zweite Metall gänzlich oder teilweise mitextrahiert worden ist. Im untersuchten System kann man nach der Extraktion aus Ammoniaklösungen eine selektive Reextraktion des Zinks (oder im Falle des Solvens D 2 EHPA – des Kupfers) mit Schwefelsäurelösungen bei Einstellung entsprechender pH-Werte im Gleichgewicht erreichen. Grundparameter sind hier der pH-Wert der wäßrigen Phase und auch die Extraktionsmittelbeladung.

Es hat sich z.B. herausgestellt, daß der laut Tab. 1 nichtselektive Extraktionsprozeß bei Verwendung von LIX 34 oder Kelex 100 durch die bevorzugte Reextraktion des Zinks in einem ziemlich breiten pH-Bereich wesentlich selektiver gestaltet werden kann (Abb. 3 und 4). Legt man für eine praktisch ausreichende Selektivität die Größen $a = b = c = 2$ fest, dann wird dieser GS_{222}-Wert in einem ziemlich breiten pH-Intervall von 1,85 (LIX 34) bzw. 2,8 (Kelex 100) realisiert. Maximale Selektivitäten werden bei pH-Werten der Reextraktionslösung von 3,5–3,6 erzielt. (An diesen sowie den folgenden Bildern ist die Abhängigkeit des Zinkextraktionskoeffizienten E_{Zn} vom pH-Wert nur der Übersicht halber angegeben; entscheidend für die Bewertung der Selektivität sind die übrigen Kurven *a*, *b* und *c*).

Je breiter der *GS*-Bereich ist, desto besser ist die erreichbare Trennung. Man sieht, daß im Falle von Kelex 100 oder LIX 34 nach der gemeinsamen Cu- und Zn-Extraktion aus ammonialkalischer Lösung und nachfolgender selektiver Reextraktion des Zn die gleichen Trenneffekte erzielt werden können wie bei Verwendung der von vornherein selektiveren Extraktionsmittel Acorga P-17, LIX 64 oder SME 529. Jedoch ist, wie aus den Abb. 5 und 6 her-

Tab. 1. Die Selektivität der Cu/Zn-Trennung bei der Extraktion aus ammoniakalischer Lösung

Extraktionsmittel	$a = \lg E_{Cu}$	$b = \lg (Cu/Zn)_0$	$c = \lg (Zn/Cu)_W$	pH-Bereich
Kelex 100	3,7–4,0	0,05	2,5–3,0	8,45–8,7
LIX 34	3,5–3,8	–0,05	1,4–2,0	8,6–9,25
Acorga P-5100	2,1–2,2	0,10–0,14	1,4–1,6	8,7–9,05
Acorga P-5300	2,6–2,75	0,12–0,14	2,0–2,1	8,65–8,8
LIX 64 N	3,0–3,75	1,2–0,9	2,8–3,6	8,7–9,35
SME 529	3,4–3,6	1,0–0,85	3,4–3,6	8,6–9,25

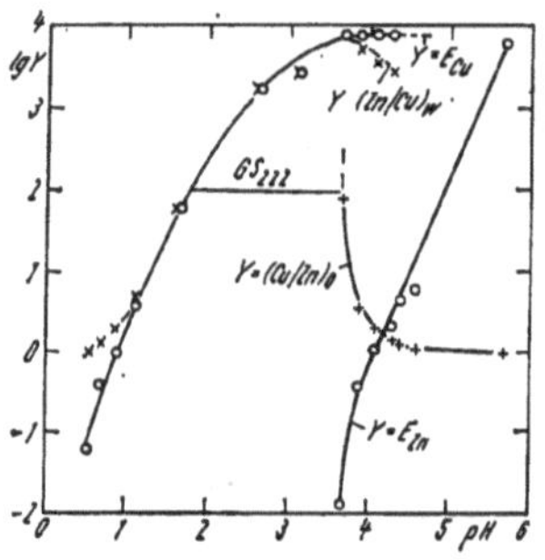

Abb. 3

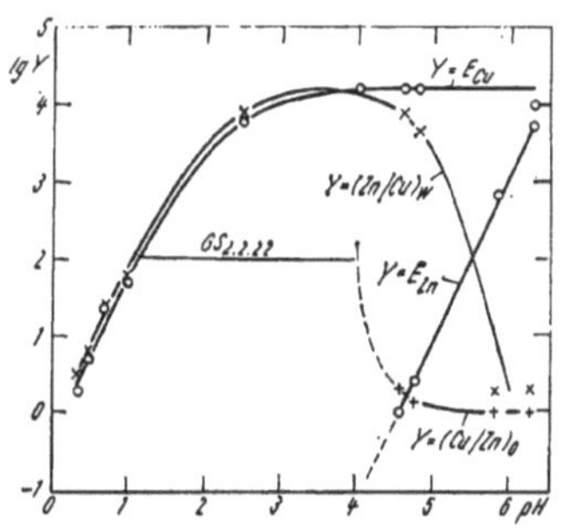

Abb. 4

Abb. 3. Die Abhängigkeit der Selektivität der Cu/Zn-Trennung vom pH-Wert bei der Reextraktion des beladenen Solvens mit Schwefelsäure. Solvens: 0,2 M LIX 34, Anfangsgehalte: 0,008 68 M Cu, 0,008 38 M Zn (beladen aus ammoniakalischer Carbonatlösung).

$GS_{2;2;2}$ pH = 1,8 – 3,65

$MS_{3,9;3,9;3,9}$; pH = 3,6

Abb. 4. Die Abhängigkeit der Selektivität der Cu/Zn-Trennung vom pH-Wert bei der Reextraktion des beladenen Solvens mit Schwefelsäure. Solvens: 0,2 M Kelex 100 + 10 % n-Dodekanol. Anfangsgehalte: 0,009 67 M Cu; 0,0971 M Zn.

$GS_{2,2;2;2}$: pH = 1,2 – 4

$MS_{4,2;4,4;4,2}$: pH = 3,5

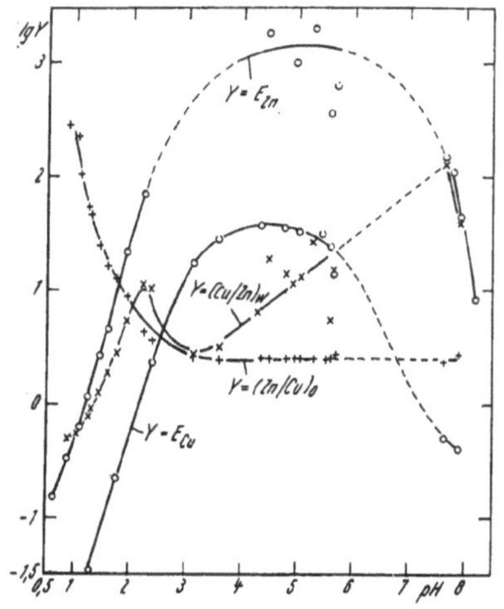

Abb. 7. Die Abhängigkeit der Selektivität der Cu- und Zn-Verteilung zwischen 1 M D 2 EHPA und Schwefelsäure vom pH-Wert bei der Reextraktion (pH ≦ 7) oder bei der Extraktion aus Ammoniak-Carbonat-Lösung. Anfangskonzentrationen im Extrakt: Cu 0,003 16 M, Zn 0,0078 M; o/w = 0,85. Anfangskonzentrationen in der wäßrigen Phase vor der Extraktion: Cu 0,01 M, Zn 0,01 M im 1 M $(NH_4)_2CO_3$; o/w = 1,45.

$MS_{1,4;0,9;0,9}$: pH = 2

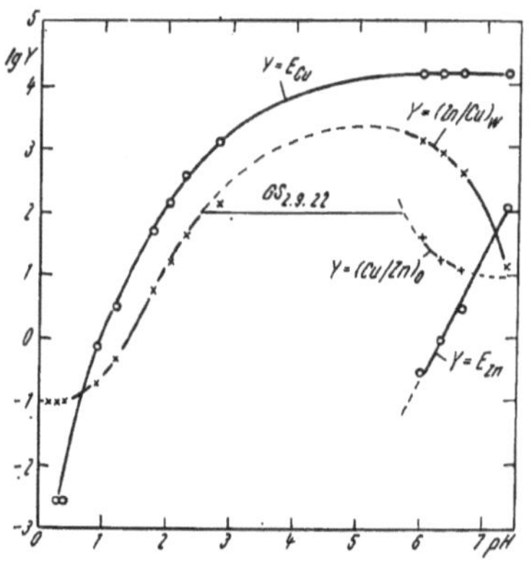

Abb. 5. Die Abhängigkeit der Selektivität der Cu/Zn-Trennung vom pH-Wert bei der Reextraktion des beladenen Solvens mit Schwefelsäure. Solvens: 0,2 M Acorga P-17 Anfangsgehalte: 0,0093 M Cu; 0,0010 M Zn

$GS_{2,9;2,2}$: pH = 2,6 – 5,7

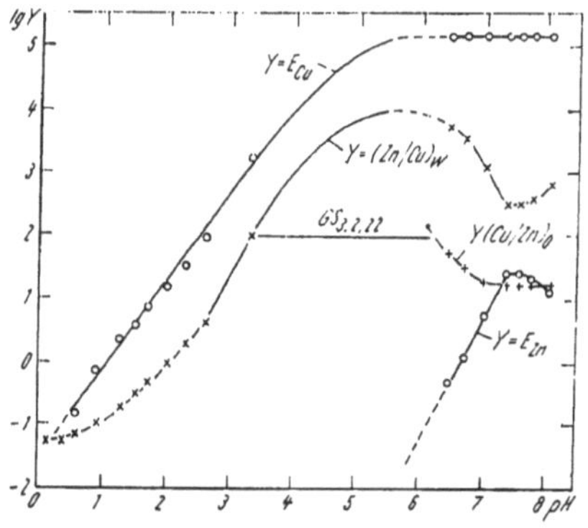

Abb. 8. Die Abhängigkeit der Selektivität der Cu/Zn-Trennung vom pH-Wert bei der Reextraktion des beladenen Solvens mit Schwefelsäure. Solvens: 0,2 M SME 529 Anfangsgehalte: 0,101 M Cu; 0,005 78 M Zn.

$GS_{3,2;2,2}$: pH = 3,35 – 6,2

$MS_{5,1;2,8;3,9}$: pH = 5,57

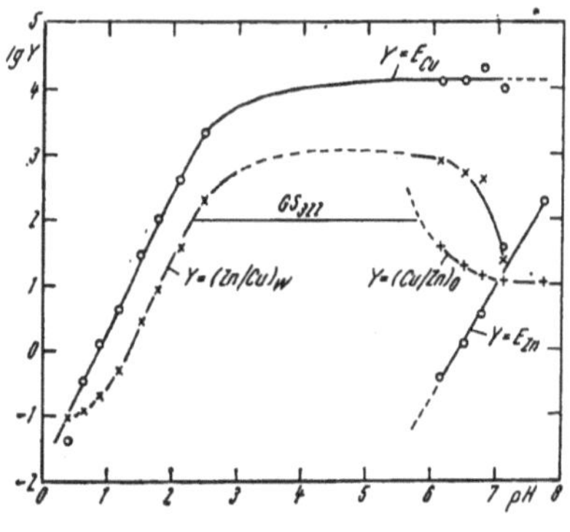

Abb. 6. Die Abhängigkeit der Selektivität der Cu/Zn-Trennung vom pH-Wert bei der Reextraktion des beladenen Solvens mit Schwefelsäure. Solvens: 0,2 M SME 529 Anfangsgehalte: 0,01 M Cu; 0,000 891 M Zn

$GS_{3;2;2}$: pH = 2,3 – 5,8

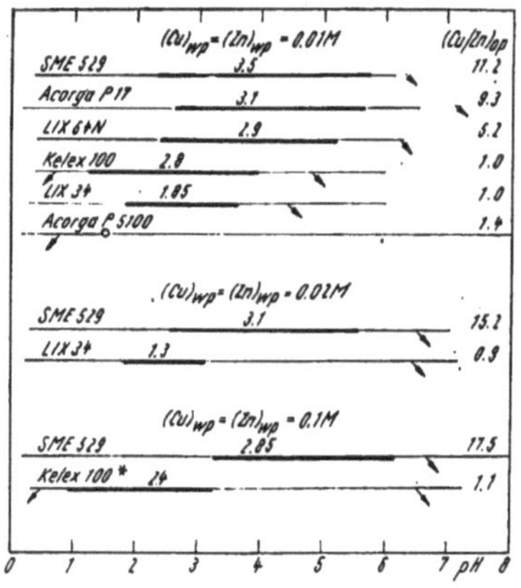

Abb. 9. Abhängigkeit der Reextraktionsselektivität vom pH-Wert für verschiedene 0,2 M Extraktionsmittel (Erläuterung der Zeichen siehe im Text).

*: Kelex 100, Konzentration 0,67 M

vorgeht, der *GS*-Bereich in den letzteren Fällen deutlich breiter. Die erreichbare Selektivität ergibt sich demnach als Summe der Trenneffekte während der Extraktion und Reextraktion.

Wie auf S. 289 erwähnt, extrahiert das Solvens D 2 EHPA Zink besser als Kupfer. Durch eine geeignete Reextraktion kann, wie Abb. 7 zeigt, die Trennung weiter verbessert werden. Zwar existiert hier kein deutlicher *GS*-Bereich mit großen *a*-, *b*-, *c*-Werten. Legt man aber eine Ordinatenparallele durch pH = 2, dann kann man aus den Schnittpunkten mit den Kurven *a*, *b* und *c* eine maximale Selektivität $MS_{1,4;0,9;0,9}$ entnehmen. In diesem Punkt beträgt die Extraktionsausbeute des Zn 95% und das Verhältnis von Zink und Kupfer in der organischen Phase ist mit ca. 8 mehr als dreimal so hoch als vor dem Strippen mit Schwefelsäure. Dennoch ist die Cu/Zn-Trennung mit D 2 EHPA weniger effektiv als die mittels Extraktionsmittel, die Cu-selektiv sind.

In dem Maße, wie das Solvens mit Metall beladen wird, verschwinden für die Solventien Acorga P-17 oder LIX 64 N die GS_{222}-Bereiche. Dagegen bleibt im Falle von SME 529, wie Abb. 8 zeigt, ein $GS_{3,2;2,2}$-Bereich in einem pH-Intervall von 2,85 auch bei hoher Beladung erhalten.

In der Abb. 9 ist die Abhängigkeit der Reextraktionsselektivität vom pH-Wert für verschiedene Extraktionsmittel dargestellt. Der stark ausgezogene Geradenteil entspricht einem $GS_{2;2;2}$-Abschnitt, der dünn ausgezogene Teil dem Bereich genügender Phasentrennung, während mit kleinen Pfeilen der Bereich schneller Phasentrennung begrenzt ist. Es ist ersichtlich, daß die $GS_{2;2;2}$-Bereiche stets innerhalb dieser Markierungen liegen, d. h. die Systeme besitzen ein gutes Phasenverhalten.

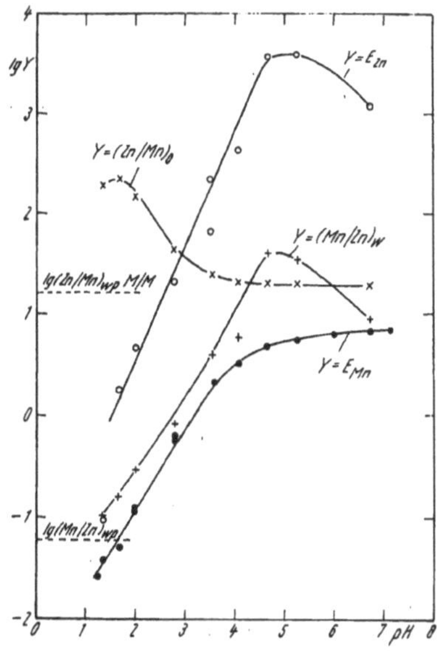

Abb. 10. Der Einfluß des pH-Wertes auf die Selektivität der Zn- und Mn-Extraktion aus Schwefelsäurelösungen mit 1 M D 2 EHPA. Anfangskonzentrationen in der wäßrigen Phase [g/dm³]: Zn 15,9; Mn 0,89; Mg 24,3. o/w = 1

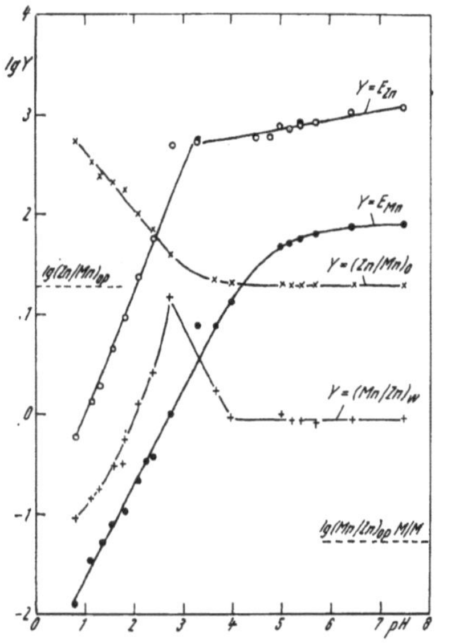

Abb. 11. Die Abhängigkeit der Zn/Mn-Trennung vom pH-Wert bei der Reextraktion des beladenen Solvens mit Schwefelsäure. Solvens: 1 M D 2 EHPA, Anfangsgehalte [g/dm³]: Zn 15,9; Mn 0,79. o/w = 1

Ein weiteres zu trennendes System sind erschöpfte Zinkelektrolyte. Diese enthalten neben Zink und Schwefelsäure auch Mangan und große Mengen Magnesium. Experimentell wurde die Trennung Zn, Mn und Mg mit dem Solvens D 2 EHPA untersucht. Zn kann bereits in einer Stufe zu 90 – 97% extrahiert werden bei gleichzeitiger wesentlicher Abreicherung des Mn. Wird hierbei eine hohe Beladung der organischen Phase mit Zn realisiert, kann das Zn/Mn-Verhältnis von ca. 15 in der Ausgangslösung bis auf 460 im Solvens gesteigert werden. Allerdings wächst dann auch die Viskosität dieser Phase stark an. Das Extraktionsgleichgewicht wird nach wenigen Minuten erreicht; Temperaturerhöhung verbessert die Zinkextraktion.

Die Abhängigkeit der Zn/Mn-Trennung vom pH-Wert bei der Extraktion geht aus Abb. 10 hervor. Die beste Selektivität wird im pH-Bereich 2 – 2,8 erreicht. Andererseits kann man auch bei pH = 7 das Zink vollständig zusammen mit 90% des Mn extrahieren und anschließend durch Reextraktion mit verdünnter Schwefelsäure (pH = 2,4) 70 – 85% des Mn entfernen. Man erhält ein Zink mit 0,03% Mn (in der Ausgangslösung betrug der Mn-Gehalt 5,6%).

Die Abhängigkeit der Selektivität der Reextraktion vom pH-Wert zeigt Abb. 11. Die Reextraktion ist in diesem Falle weniger selektiv als die Extraktion. Das anwesende Magnesium wird gewöhnlich nur wenig (≦ 2%) mitextrahiert.

Hier wurde stets nur einstufig gearbeitet, aber es ist bekannt, daß der Trenneffekt durch mehrstufiges Arbeiten im Gegenstrom wesentlich verbessert werden kann.

Eingegangen am 2. 12. 1985

Isotopenpraxis 22 (1986) 8, S. 292 bis 293

Kurze Originalmitteilung

The possibility of rapid assessment of the degree of internal contamination with ^{131}I by nose blows activity measurement

J. Popper, Z. Hölgye

(Institute of Industrial Hygiene in Uranium Industry)[1]

The degree of internal contamination of persons with radioactive substances can be generally determined by whole-body measurement and from activity excreted in urine and feces. Activity measurement of nose blows for the assessment of internal contamination degree was used only uniquely [1]. The activity measurement of nose blows is, however, used for individual nomitoring on plutonium workplaces [2] and as a criterion for the classification of persons according to contamination relevance at radiation accidents [3]. From published data, it follows that there is no important difference between the intake determined from nose blows activity and that determined by whole-body measurement or derived from activity of excreta.

Within the framework of regular monitoring of the degree of personnel internal contamination in radiopharmaceuticals production of natrium o-iodohippurate and bengal rose in Nuclear Research Institute, Řež, the internal contamination by radioiodine was found only sporadicly [4]. Individual monitoring was performed by whole-body measurement, by the direct measurement of ^{131}I in thyroid gland and by ^{131}I determination in urine. As it was proved that internal contamination is caused by inhaling [5], we tried in several cases to determine ^{131}I intake by activity measurement of nose blows under parallel determinations of ^{131}I in 24 hours urine.

The aim of this work is a summarization of present experience and a comparison with results of nose blows activity measurement from the point of a possibility of the use for rapid assessment of internal contamination degree at exceptional events and radiation accidents.

Procedures

Nose blows were taken at individual workers by wadding tampons wetted before use by physiologic solution. A new tampon was used for each nostril. Nose blows were taken always immediately following the most hazardous operation (from the point of internal contamination) in the process of radiopharmaceuticals production. This phase of production was established on the basis of detailed analysis of the production cycle [4]. ^{131}I activity in nose blows was measured by low background laboratory unit NRB-213 [Tesla].

Simultaneously with taking of nose blows, the samples of urine from following 24 hours were collected for radiochemical determination of excreted ^{131}I. The ^{131}I determination in urine was accomplished by a method based on the extraction of iodine after preceding mineralization of urine [6]. On the basis of results of ^{131}I determination in urine, the intakes were calculated using the function recommended by the ICRP publication No 30. [7]. The relationship between the activity of ^{131}I determined in nose blows A_{nv} [Bq] and the intake P [Bq] derived from excretion by urine was presented graphically. The degree of correlation of both variables was expressed as the correlation coefficient r_k according to the relation

$$r_k = \frac{\sum (x - \bar{x}) \cdot (y - \bar{y})}{\sqrt{\sum (x - \bar{x})^2 \cdot (y - \bar{y})^2}}$$

For practical use, the ratio A_{nv}/P for every individual and the mean value of this ratio for all individuals monitored was calculated.

Results and discussion

The relationship between ^{131}I activity in nose blows and ^{131}I excretion by urine for personnel in the production of radiopharmaceuticals labelled with ^{131}I was studied in 48 cases in the group of six individuals. The relationship of both variables is graphically presented in Fig. 1. It can be seen that there is a very good correlation what is also confirmed by the value of correlation factor $r_k = 0{\cdot}938$. The r_k value was calculated only from results obtained from nose blows samples taken immediately (< 2 min) following work. If taken later, the value of ratio A_{nv}/P significantly decreased. E.g. for samples taken 3 hours following work, the ratio A_{nv}/P was one order lower. ^{131}I activity in nose blows varied in the range of $0{\cdot}1 - 5{\cdot}0$ Bq. The ratio A_{nv}/P covered range of $(2{\cdot}7$ to $16{\cdot}5) \cdot 10^{-3}$ with the mean value $9{\cdot}8 \cdot 10^{-3}$. That means that under given conditions, the intakes of ^{131}I were on the average 100 times higher than ^{131}I content in nose blows. Similar results obtained *Barall* [3] who studied ^{131}I activity in nose blows, thyroid gland and urine upon collective contamination

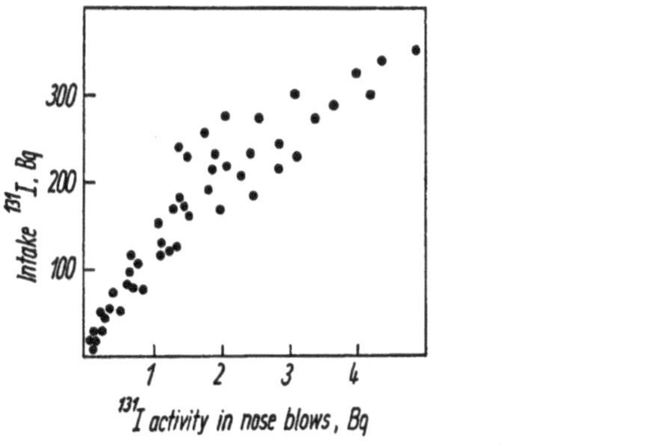

Fig. 1. The relation of ^{131}I activity in nose blows to ^{131}I intake

[1]) Address: Příbram-Kamenná, PO-Box 262 31, Czechoslovakia

of personnel. The possibility of secondary samples contamination from lips, nose and face, that was adviced upon by *Jech* et al., could be excluded since surface contamination of examined individuals was not found [5]. Similarly, the contamination by other radionuclides does not come into consideration as solely radionuclide ^{131}I of high radiochemical purity was used for production.

Conclusion

The correlation, established between ^{131}I activity in nose blows and intakes derived from excretion by urine, gives possibility of rapid and relatively precise assessment of the relevance of internal contamination. Of great significance is, in the first place, the promptitude of intake detection with regard to the necessity of rapid classification of personnel according to the degree of internal contamination at radiation accidents. Using the ratio A_{nv}/P, the intake of ^{131}I can be determined by activity measurement of nose blows within 20 minutes after samples off-take.

Received November 14, 1985

References

[1] *B. Holliday, G. W. Dolphin, H. J. Dunster,* Health Phys. 18 (1970) 529

[2] *J. Jech, K. R. Heid, H. V. Larson,* Handling of radiation accidents, Proceedings of a symposium organized by the IAEA in collaboration with the WHO and held in Vienna 19 – 23 May, 1969, p. 77 – 93

[3] *R. C. Barall,* Health physics society 8th annual meeting, New York June 1963

[4] *J. Chyský, A. Garba, Z. Hölgye, J. Popper,* Závěrečná zpráva úkolu P17-335-236-01/03, ÚHP UP Příbram, pob. Řež, Řež u Prahy, 1980

[5] *J. Popper, Z. Hölgye,* Pracov. Lék. 35 (1983) 410

[6] Informační zprávy z oboru hygieny práce a nemocí z povolání, příloha č. 3, ÚHPCHP, Praha 1967

[7] ICRP publication No. 30, Limits for intakes of radionuclides by workers, Pergamon press, Oxford 1979

Isotopenpraxis 22 (1986) 8, S. 293 bis 296

Buchbesprechungen

Optoelektronische Halbleiterbauelemente

R. Paul, Stuttgart: Teubner 1985. 300 S., 141 Abb., 20 Tab., 197 Lit. (Teubner-Studienskripten, Angewandte Physik/Elektrotechnik, Band 96); kart.

Das vorliegende Buch ist aus Skripten einer Einführungsvorlesung über Halbleiterbauelemente für Studenten der Elektrotechnik entstanden. Es bietet einen Überblick über die Wirkprinzipien, Eigenschaften und Anwendungsmöglichkeiten moderner optoelektronischer Halbleiterbauelemente, d. h. Lumineszenz- und Laserdioden, Displays, Fotoempfänger, Bildaufnahmeeinheiten, Optokoppler und optische Übertragungssysteme. Vom Leser werden Grundkenntnisse in Physik und Elektronik vorausgesetzt, wie sie in den Einführungskursen der technischen Hoch- und Fachschulen vermittelt werden.

Das Buch ist leichtverständlich und praxisbezogen geschrieben. Es ist zum Gebrauch neben Vorlesungen sowie für das Selbststudium vorgesehen. Die gute Gliederung des Stoffes, das reichhaltige Skizzenmaterial sowie ein ausführliches Stichwortverzeichnis sind besondere Vorzüge des Buches.

Im ersten Abschnitt wird die Phänomenologie der Strahlung und ihrer Wechselwirkung mit dem Festkörper an Hand des Bändermodells behandelt. Der Autor geht übersichtsartig auch auf die Funktion von flachen Störstellen, isoelektrischen und tiefen Zentren ein. Darüber hinaus enthält der Abschnitt eine übersichtliche Zusammenfassung der Definitionen von Größen zur Kennzeichnung von Strahlung. Der zweite Abschnitt ist den Halbleiterstrahlungsquellen gewidmet. Es werden die Wirkprinzipien, einige Eigenschaften und Kennwerte von LEDs mit einfachen pn-Übergängen sowie mit Hetero-, Schottky- und MIS-Übergängen beschrieben. Außerdem werden das Frequenz- und Modulationsverhalten der LEDs sowie die gebräuchlichsten Materialien, Bauformen und Grundschaltungen charakterisiert. Die wichtigsten Displaytechniken werden miteinander verglichen, LED- und Flüssigkeitsanzeigen sind ausführlich beschrieben. Die Erläuterungen zu den verbreiteten Laserstrukturen (SH, DH, LOC, BH, TIS, CSP u. a.) werden durch informatives Bildmaterial ergänzt. Der dritte Abschnitt führt in das weite Gebiet der Strahlungsdetektoren des optischen Bereiches ein. Röntgen- und Kernstrahlungsdetektoren werden nicht betrachtet. Der Autor behandelt Fotoleiter, -dioden, -transistoren, -thyristoren, Halbleiterfotokatoden, Solarzellen und integrierte Bildsensoren. Ein Unterabschnitt bietet eine zusammenfassende Darstellung zum Rauschen von Fotodetektoren. Der Autor kann selbstverständlich in dem Rahmen des Buches keine vollständige Aufzählung der für Strahlensensoren geeigneten Halbleitermaterialien liefern, er gibt aber einen guten Einblick in die Vielzahl der Entwicklungsrichtungen auf diesem Gebiet. In den letzten beiden Abschnitten der Buches diskutiert der Autor die Besonderheiten des Einsatzes von Strahlungsempfängern und -sendern in Optokopplern und optischen Übertragungssystemen. Es werden die wichtigsten Begriffe erklärt, Grundschaltungen vorgestellt und einige Angaben zur Entwicklung der Wellenleitertechnik gemacht.

Das Buch darf als eine gute Einführung für das sehr komplexe Gebiet der optoelektronischen Halbleiterbauelemente empfohlen werden. Es besticht durch seine Übersichtlichkeit und die knappe klare Darstellung der technischen Zusammenhänge. Von großem Wert sind die Hinweise auf weiterführende Literatur; darunter sind 30 Bücher (hauptsächlich aus den 80er Jahren), die das Gesamtthema betreffen, und etwa 170 Literaturzitate zum Stoff der einzelnen Abschnitte (40% davon sind 1980 oder später erschienen).

A. Blüschke

Atomphysik, Eine Einführung

T. Mayer-Kuckuk. Stuttgart: B. G. Teubner 1985. 3. Auflage, 250 S., 119 Abb., 7 Tab. (Teubner Studienbücher Physik), kartoniert DM 32,–

Der Inhalt des Buches entspricht einer einsemestrigen Einführungsvorlesung in die Atomphysik und ist deshalb bewußt elementar gehalten. Der Leser (Hörer) soll erkennen, welche Gesetzmäßigkeiten in der Physik der Atome herrschen und wie diese Gesetzmäßigkeiten in mathematischen Formeln dargestellt werden. Das Buch ist in insgesamt 11 Kapitel eingeteilt, die in sich überschaubar dem Studenten Teilgebiete der Atomphysik, wie Teilchen und Welle, Magnetfeld und Spin des Elektrons, Emission von Lichtquanten oder kohärente und inkohärente Strahlungsquellen — um nur einige zu nennen —, nahebringen sollen. Neu zum Unterschied früherer Auflagen ist das Kapitel Ungewöhnliche Atome, in welchem das Positronium und Myonium sowie myomische und hadronische Atome behandelt werden.

Trotz der bereits vorliegenden 3. Auflage sind eine Reihe von Ungenauigkeiten bzw. Druckfehlern enthalten. Auf S. 33 werden die Materiewellen von de Broglie behandelt und mit der Jahreszahl 1927 in Verbindung gebracht. Dies ist falsch: Prinz de Broglie hat zum ersten Mal in seiner Dissertation "Recherches sur la théorie des quanta" 1924 von Materiewellen gesprochen und in den Jahren 1925 [Ann. Physik **3** (1925) 22] und 1926 [J. de Physique **1** (1926) 1] seine Ergebnisse veröffentlicht, die dann allerdings 1927 durch C. Davisson und L. H. Germer in den Bell Telephone Laboratories an Ni-Einkristallen glänzend bestätigt wurden [Phys. Rev. **30** (1927) 705; Proc. Nat. Acad. Amer. **14** (1928) 317].
Einige Vorzeichen sind falsch (S. 95, S. 140), und auf S. 242 befindet sich ein ω zuviel (oder ein = zu wenig). Die Abb. 66 auf S. 159 ist nicht das H-Atom (in der Zeichnung), sondern das He-Atom. In der Tabelle A2 auf S. 243 sollte der quantenmechanische Zustandsvektor 3 als ket-Vektor geschrieben werden. In der Fußnote auf S. 157 ist ein Ausdrucksfehler.

Obwohl der Autor sich bereits in der Einleitung für seine Verwendung des cgs-Systems entschuldigt, sollte doch angestrebt werden, den SI-Einheiten den Vorzug zu geben, insbesondere da es sich um einen Einführungskursus in Atomphysik handelt. Das angeführte Argument, man könnte den Faktor $4\pi\varepsilon_0$ im cgs-System „einsparen", ist wohl nicht überzeugend und ausreichend; denn einige Formeln sind im cgs-System wegen des Auftretens neuer Konstanten wie c und π sogar länger als im SI. Eine Atomphysik zu beschreiben, ohne eine fundierte und postulierte Quantenmechanik in Anspruch zu nehmen, ist gewiß schwierig; der Autor hat es versucht. Dem Lesenden und Lernenden wird aber ein guter Einstieg in dieses Gebiet geboten, der ihm bei einer weiterführenden und vertiefenden Beschäftigung bestimmt als gute Grundlage dienen wird.

H. Hübner

Die nukleare Nacht

P. R. Ehrlich, C. Sagan, L. Thomas, P. J. Crutzen u. a., Übers. H. Kober. Köln: Kiepenheuer & Witsch 1983. 320 S.; DM 19,80

„Wenn es unserer Welt vergönnt sein sollte, ihre Geschichte fortzusetzen, könnten sich die in diesem Buch dargelegten wissenschaftlichen Entdeckungen eines Tages als die wichtigsten Forschungsergebnisse erweisen, die in der langen Geschichte der Wissenschaft jemals gefunden wurden." So beginnt das Vorwort von L. Thomas, einer der 28 Wissenschaftler aus Ost und West, die an der Ausarbeitung der klimatisch, genetischen und biologischen Auswirkungen nuklearer Kriege beteiligt waren, die in diesem Buch beschrieben sind. Das Buch knüpft an die Ergebnisse von Untersuchungen an, die vom Herausgeber der Zeitschrift Ambio 1981 initiiert worden sind [Ambio **11** (1982) Nr. 2/3]. Danach würde in einem globalen Kernwaffenkrieg ein Drittel bis die Hälfte der Menschheit durch die von nuklearen Explosionen ausgehenden Druck- und Hitzewellen, durch die Sofortstrahlung, den radioaktiven Fallout und deren kurzfristigen Folgen umkommen. Inzwischen hat sich jedoch gezeigt, daß nukleare Kriege drastische Veränderungen in der Umwelt und in den Ökosystemen auslösen würden, deren Folgen für die Menschheit von noch größerem Ausmaß wären als die bis 1982 untersuchten, und zwar selbst dann, wenn man nicht von mehr als 100 Explosionen über großen Städten oder 2000 bis 3000 Explosionen größerer Sprengkraft in Bodennähe zur Bekämpfung militärischer Ziele ausgeht, also einem Ausmaß des Einsatzes nuklearer Waffen, das um den Faktor 150 bzw. um den Faktor 6 unter den durch die vorhandenen Kernwaffenarsenale gegebenen Möglichkeiten liegt. Danach würde der größte Teil der Menschheit innerhalb von wenigen Monaten bis zu einigen Jahren umkommen. Die wenigen Überlebenden wären, auf ozeanischen Inseln und in Küstenregionen zu einem Dasein von Jägern und Sammlern verurteilt, einer unwirtlichen Umwelt ausgesetzt, die ihnen unbekannt ist, an die sie nicht angepaßt sind und in der sie innerhalb einer längeren Frist wahrscheinlich aussterben würden.

Daß nach diesen neuesten Erkenntnissen eben nicht nur die Erhaltung der Zivilisation, sondern sogar die Erhaltung der Menschheit auf dem Spiele steht, daß dies (von einer bestimmten, überraschend niedrigen Schwelle an) auch für sogenannte begrenzte Atomkriege zutrifft und daß Angreifer, Angegriffene und Unbeteiligte von den ökologischen Auswirkungen in gleichem Maße betroffen wären, liegt in erster Linie daran, daß bei den gewaltigen Flächenbränden, die Kernwaffenexplosionen auslösen würden, große Mengen Rauch und Ruß in die Troposphäre und Stratosphäre geschleudert würden, die (wegen Reflexion und Absorption eines großen Teils des Sonnenlichtes) auf der Erdoberfläche anhaltende Dunkelheit und Kälte verursachen würden in einem Ausmaß, daß die landwirtschaftliche Produktion zum Erliegen käme. Hinzu käme ein stärkerer Fallout in mittleren räumlichen und zeitlichen Entfernungen von den Explosionsorten, als das bisher angenommen wurde, die biotische Verarmung der Tier- und Pflanzenwelt unter dem Einfluß der ionisierenden Strahlung, die mit einem Hervortreten der „schädlichen" Tiere und Pflanzen einhergehen würde, die Bildung giftiger gasförmiger Verbrennungsprodukte bei Großbränden vor allem in städtischen Gebieten und die Erhöhung der Intensität im UV-B des Sonnenlichtes nach der nuklearen Nacht infolge des Ozonabbaus in der Atmosphäre unter dem Einfluß der bei den Explosionen gebildeten nitrosen Gase.

Das Buch, dessen Inhalt in seinen wesentlichen Teilen von allen Menschen erfaßt werden kann, die über das von den polytechnischen Oberschulen der DDR vermittelte naturwissenschaftliche Wissen verfügen, präsentiert die Ergebnisse einer 1983 in Washington veranstalteten Tagung einschließlich der dort geführten Diskussionen, den Verlauf einer Konferenzschaltung von der Tagung nach Moskau sowie im Anhang zwei einschlägige Publikationen in der Science. Der Verlauf der Konferenz zeigt einen ebenso erfreulichen wie erstaunlichen, hohen Grad der Übereinstimmung der über 100 Teilnehmer in den wissenschaftlichen Auffassungen, auch hinsichtlich der hohen Sicherheit in den in dieser Rezension zusammengefaßten Aussagen über die Konsequenzen eines Kernwaffenkrieges für die Menschheit. Die in die vorgestellten Modelle eingehenden Daten über die nuklearen Arsenale der UdSSR und der USA geben übrigens nicht den geringsten Anhaltspunkt für die von der USA-Regierung behauptete nukleare Überlegenheit der UdSSR, mit der die Raketenstationierung in Westeuropa begründet wurde.

Der Vorsatz der Veranstalter, politische Fragen auszuschließen, konnte aus allzu verständlichen Gründen nicht durchgesetzt werden. Dennoch gab es wenigstens in der entscheidenden Frage Übereinstimmung. So führte *J. Welichow*, Vizepräsident der AdW der UdSSR, aus: „... der einzig mögliche Schluß lautet, daß unsere Kernwaffen nicht als Werkzeuge des Krieges und auch nicht als Werkzeuge der Politik benutzt werden können. Sie sind lediglich Werkzeuge des Selbstmords" (S. 231), während *C. Sagan*, Direktor des Laboratoriums für planetarische Studien an der Cornell University angesichts der niedrigen Schwelle der katastrophalen Auswirkungen auf die Umwelt erklärte: „Die Wahrscheinlichkeit ist groß, daß Land A, wenn es Land B mit einem erfolgreichen Erstschlag überfällt, der nur „harten" militärischen Zielen gilt, dadurch Selbstmord begangen hat, auch wenn Land B nicht einen Finger zur Vergeltung gerührt hat" (S. 91). Selbst dem mystischen Vorwort zur deutschsprachigen Ausgabe läßt sich eine Aussage abgewinnen, der man gern zustimmt: „Ein Buch der Hoffnung. Es wird der Friedensbewegung Auftrieb geben".

Der Rezensent möchte hinzufügen: Mögen immer mehr Menschen dieser am Rande eines Abgrundes wandelnden Welt Einsicht in die sozialen Wurzeln des Krieges erlangen und ihr Denken und Handeln davon ableiten, von welcher Gesellschaftsordnung solche konstruktiven Abrüstungsvorschläge ausgehen, wie der vom 15. 1. 1986 über die Beseitigung nuklearer Waffen bis zum Ende dieses Jahrhunderts, und in welcher sozialen Ordnung diejenigen Politiker in ihre Ämter gewählt werden, die von der Begrenzbarkeit und Gewinnbarkeit nuklearer Kriege, von nuklearen Erstschlagsstrategien und von der Militarisierung des Weltraumes träumen. Möge das Buch, dessen Inhalt Allgemeingut der Menschheit werden muß, dazu beitragen, daß die Verhandlungen über die Begrenzung des Wettrüstens und seine Ablösung durch wirksame Abrüstungsmaßnahmen Erfolge zeitigen.

K. Welzel

Food Irradiation Processing

Proc. of a Symp., Washington, D.C., March 4–8, 1985. Vienna: IAEA 1985. 554 S., div. Abb. u. Tab. (Series-STI/PUB/695); ÖS 1100,–

Der Band enthält die Vorträge sowie Kurzfassungen der Poster eines gemeinsam von der IAEA und der FAO organisierten Symposiums über die Lebensmittelbestrahlung. Gegenüber früheren, in der gleichen Reihe erschienenen Fortschrittsberichten zu diesem Thema ist festzustellen, daß in vielen Artikeln über die Entwicklung technologischer Lösungen für die Lebensmittelbestrahlung und die Markteinführung bestrahlter Produkte berichtet wird. Die Umsetzung der auf dem Gebiet der Lebensmittelbestrahlung vorliegenden Erkenntnisse in die Praxis und die Förderung ihrer Nutzung in der Lebensmittelindustrie war auch das erklärte Ziel der Tagung zu deren Abschluß ein Expertenausschuß unter Vorsitz von *J. S. Sivinski* (Albuquerque, USA) empfahl, dem Nachweis einer ökonomisch tragfähigen Realisierbarkeit, der Schaffung der gesetzlichen Grundlagen für die Lebensmittelbestrahlung und der Verbraucheraufklärung besondere Aufmerksamkeit zu widmen. Vorteile aus der Nutzung von Verfahren der Lebensmitelbestrahlung ergaben sich vor allem durch die Reduzierung von Verlusten während der Lagerung landwirtschaftlicher Produkte, durch die Förderung des internationalen Handels (da der Verderb von Produkten drastisch reduziert wird und nationale Hygienerichtlinien für pflanzliche und tierische Produkte durch Inaktivierung pathogener Keime oder von Schadinsekten erfüllt werden können) und durch Beiträge zur Verbesserung der Volksgesundheit, da durch Lebensmittel übertragene infektiöse Erkrankungen vermieden werden können.

In mehreren Artikeln wird die gegenwärtige Entwicklung der Lebensmittelbestrahlung in Europa und im Mittleren Osten (*J. Farkas*; Budapest), in lateinamerikanischen Ländern (*T. Rubio*; Santiago de Chile), in Afrika (*B. Chinsman*; Dakar) und in Asien (*P. Loaharanu*; IAEA Wien) dargestellt.

In weiteren Arbeiten werden technologische und ökonomische Probleme der Strahlenbehandlung von Speisezwiebeln, Kartoffeln und tropischen Früchten, der Insektenbekämpfung in Hülsenfrüchten und Getreide, der Salmonellenbekämpfung im Fleisch und Geflügel sowie der Keimzahlreduzierung in Trockenprodukten wie Gewürze, Trockengemüse und Gelatine diskutiert. Pilotanlagen sowie Projekte für multivalent nutzbare und zu speziellen Zwekken einsetzbare Bestrahlungsanlagen werden vorgestellt.

Der vorliegende Band aus der Serie der IAEA-Fortschrittsberichte gibt viele Anregungen und Hinweise für Einsatzmöglichkeiten dieses (schon 40 Jahre bekannten) Verfahrens der Lebensmittelbehandlung und sollte vor allem von der Lebensmittelindustrie, den Hygiene- und Veterinärhygiene-Einrichtungen zur Kenntnis genommen werden. Hilfe für Verbraucheraufklärung gibt es nicht, da die gesundheitliche Unbedenklichkeit des Verfahrens als erwiesen und infolge früherer Publikationen in der gleichen Reihe als bekannt vorausgesetzt wird.

G. Hübner

Ion-Selective Electrodes, 4
Fourth Symposium held at Mátrafüred, Hungary, October 8–12, 1984

Hrsg.: E. Pungor u. I. Buzás. Budapest; Akadémiai Kiadó 1985. XVI + 766 S., zahlr. Abb. u. Tab.; geb. M 233,50

Das vorliegende Buch enthält die Beiträge des 4. Symposiums über ionensensitive Elektroden, das im Oktober 1984 in Mátrafüred stattfand und an dem 89 Wissenschaftler aus 18 Ländern teilnahmen. Dieser Band erscheint gleichzeitig bei Elsevier Science Publishers (Amsterdam) in der Reihe „Analytical Chemistry Symposia Series".

In dem umfangreichen Band sind vier Plenarvorträge (110 Seiten), acht Hauptvorträge (169 Seiten) und 39 Diskussionsvorträge (438 Seiten) abgedruckt sowie auf weiteren 30 Seiten zwei Beiträge zu einem Rundtischgespräch über Probleme der Standards beim Arbeiten mit ionensensitiven Elektroden. Die Plenarvorträge beinhalten eine Übersicht über Mechanismen chemischer Sensoren aus elektrochemischer Sicht von R. P. Buck (Chapel Hill), einen Beitrag von K. Cammann (Ulm) und G. A. Rechnitz (Newark) über Polarisationsuntersuchungen an ionensensitiven Elektroden, in dem die Autoren ein einfaches „Mischpotential-Modell" für ionensensitive Membranen vorschlagen, weiterhin einen Vortrag von N. Ishibashi (Fukuoka) über Selektivitäten von Flüssigmembranelektroden und Möglichkeiten zur Selektivitätssteigerung sowie Ausführungen von W. Simon (Zürich) über klinisch-chemische Anwendungen und daraus resultierende Anforderungen an die Membranelektroden. Bemerkenswerte Hauptvorträge sind u. a. eine kurze Übersicht über modifizierte Elektroden und deren analytische Anwendungen (R. A. Durst, Washington) und Beiträge von J. Janata (Salt Lake City) and Yu. G. Vlasov (Leningrad) über ionensensitive Feldeffekttransistoren, die sogenannten ChemFETs bzw. ISFETs. Im übrigen dominieren analytische Anwendungen ionensensitiver Elektroden (hauptsächlich zu Untersuchungen von Komplexgleichgewichten und umweltrelevanten Proben), Arbeiten über verschiedenste Membranmaterialien und -typen, die mit elektrochemischen und anderen physikalisch-chemischen Methoden untersucht wurden, sowie Beiträge zu Interferenzen, die die Anwendbarkeit ionensensitiver Elektroden limitieren, bzw. ihrer Unterdrückung.

Insgesamt gibt der Band einen guten Überblick über ionensensitive Elektroden, ihre Anwendungsmöglichkeiten und auch über angrenzende Gebiete. Die Literaturhinweise sind umfangreich und recht aktuell, etwa 50% der insgesamt ca. 800 zitierten Publikationen sind in den achtziger Jahren erschienen. Die wissenschaftliche Qualität des Buches ist unbestritten, aber bei der Vorbereitung des Druckes und der Herstellung des Buches wäre eine größere Sorgfalt wünschenswert gewesen. So wurden z.B. Abbildungen (S. 518 u. 519) und Text (S. 110 u. 111) vertauscht, es sind viele Druckfehler vorhanden, und im Besprechungsexemplar waren zwischen den Seiten 657 und 672 acht unbedruckte Seiten enthalten.

Trotzdem ist der Band Bibliotheken, Instituten usw. als aktuelles Werk über ionensensitive Elektroden zu empfehlen. Auf Grund des Preises und der Qualitätsmängel dürfte es allerdings kaum von Wissenschaftlern für ihre Privatbibliothek gekauft werden. *K.-H. Lubert*

Integrierende Emanationsmessungen zur Suche von Rohstofflagerstätten

V. K. Titov, V. A. Venkov, T. L. Avdeeva, E. I. Kuvšinnikova. Moskau: Nedra, Leningradskoe otdelenie 1985. 132 S., 59 Abb., 14 Tab., 56 Lit. (russ.); brosch. M 2,50

Schon wenige Jahre nach der Entdeckung des radioaktiven Edelgases Radon erkannte man, daß sich verschiedene Gesteine, aber auch tektonische Störungen durch unterschiedliche Abgabe des Radons an die Boden- oder Außenluft zu erkennen geben. Damit war die Grundlage für die Entwicklung der Emanations- oder Radon-Aufnahme als Mittel zur Suche von tektonischen Störungen (Spalten, Kluftsysteme, Wasserleiter) und von Rohstoff-Lagerstätten gegeben. Dieses Verfahren wurde zu beträchtlicher technischer Vollkommenheit (Empfindlichkeit, Genauigkeit, Meßleistung) entwickelt. Aber allen Radon-Aufnahmen haftet der Mangel an, daß der Betrag der Meßgröße sehr stark von äußeren Bedingungen abhängt (Temperatur, Wind, Luftdruck und dessen Schwankungen, Durchfeuchtung des Bodens). Deshalb sind die Ergebnisse von Radon-Aufnahmen, die zu verschiedenen (Jahres-)Zeiten und bei verschiedenen meteorologischen Bedingungen ausgeführt wurden, oftmals wenig vergleichbar.

Einen grundsätzlichen Ausweg verspricht die Anwendung von integrierenden Detektoren, wie sie in den Festkörper-Spur-Detektoren oder in alpha-empfindlichen Halbleitern vorliegen. Schon kurze Zeit, nachdem in der Kernphysik diese Methoden erforscht und entwickelt worden waren, begann man auch ihre Nutzung für die geophysikalische Radon-Aufnahme zu untersuchen. Über die Ergebnisse dieser Forschungen legen die Autoren eine zusammenfassende Beschreibung vor, in der sie hauptsächlich die in der UdSSR gesammelten Erfahrungen darlegen.

Nach einer kurzen historischen Einleitung werden die physikalischen Grundlagen behandelt: das Verhalten des Radons im Boden und sein Übertritt in die zur Messung geschaffenen Hohlräume. Es folgen die ausführliche Beschreibung der Spurenmethode (track etching) und anderer integrierender Verfahren. Zum Schluß werden praktische Anwendungsbeispiele vorgeführt und Vorschläge zur Nutzung der Methode gegeben.

Das vorliegende Büchlein ist eine nützliche Zusammenfassung des erreichten Standes und kann zur raschen Einarbeitung in dieses Gebiet empfohlen werden.

K. H. Köhler

Evolution of Matter and Energy on a Cosmic and Planetary Scale

M. Taube. New York-Berlin-Heidelberg-Tokyo: Springer-Verlag 1985. 289 S., 136 Abb., 37 Tab., (engl.); kart. DM 76.–

Dieser anspruchsvolle Titel beschreibt die Entwicklung unseres Universums von den Anfängen bis in die ferne Zukunft auf einer naturwissenschaftlichen Grundlage. Es ist dabei die Absicht des Autors, die wesentlichen Abläufe in einer einfachen und verständlichen Art dem Leser nahezubringen. Die Entstehungsgeschichte des Buches reicht bis in das Ende der 50er Jahre zurück, wo der Autor Vorlesungen darüber an der Warschauer Universität gehalten hat und von 1969–1983 einen weiteren Vorlesungszyklus an der Eidgenössischen Technischen Hochschule und an der Universität in Zürich hielt.

Das Buch ist in 9 Kapitel gegliedert, beginnend mit den Elementarteilchen und ihren Wechselwirkungen in einer sehr frühen Phase unseres Universums über die kernphysikalische Entwicklung (Nukleogenese), die chemische Entwicklung mit ihren Bezügen zur Entstehung des Lebens, die Entwicklungsgeschichte unserer Erde mit den ablaufenden Stoff- und Energieströmen und endet mit den gegenwärtigen Problemen der Menschheit bis hin zur Entwicklung unseres Universums in den folgenden Jahrmillionen. Ein breites Band, das überschaubar darzustellen ist, wobei manches nur angedeutet ist. Hervorzuheben ist die an jedes Kapitel angeschlossene Frage, welche Folgerungen für die künftige Entwicklung der Menschheit aus dem Vorhergesagten gezogen werden können: wenn auch nur skizzenhaft formuliert, so doch das Wesentliche erfassend.

Neben einigen Druckfehlern wäre zu bemerken, daß manchen Zahlenangaben, wie beispielsweise bei den großen Stoffkreisläufen in der Natur, ein unkritischer Charakter anhaftet. Es sollte doch besser deutlich gemacht werden, daß sich viele Einzelheiten, so bedeutsam sie auch einzuordnen sind, noch in der Phase der Erforschung und der Einordnung in einem globalen Maßstab befinden. Ferner sind die Schraffuren bei vielen Abbildungen didaktisch nicht gut gewählt, worunter nicht nur die Übersichtlichkeit leidet. Alles in allem ein Versuch einer Gesamtschau, die einem interessierten Leserkreis – oder einer Hörerschaft in Vorlesungen – wichtige Einblicke in die modernen Ansichten der Entwicklung unseres Universums erlaubt. *H. Hübner*

ISOTOPENPRAXIS

Band 22 | 1986 | Heft 8

Contents — Inhalt

►Language of Publication — ►Publikationssprache

Isotopenpraxis is indexed in Current Contents/PC & ES, Science Citation Index, and INIS Atomindex

Kolibri 360 GmbH
Ferdinand-Jühlke-Straße 7
99095 Erfurt, Deutschland
produktsicherheit@kolibri360.de